INTRODUCTION
TO
MATERIALS
SCIENCE
AND
ENGINEERING

INTRODUCTION
TO

MATERIALS SCIENCE AND ENGINEERING

Yip-Wah Chung

CRC Press
Taylor & Francis Group
Boca Raton London New York

CRC Press is an imprint of the
Taylor & Francis Group, an informa business

CRC Press
Taylor & Francis Group
6000 Broken Sound Parkway NW, Suite 300
Boca Raton, FL 33487-2742

International Standard Book Number-10: 0-8493-9263-2 (Hardcover)
International Standard Book Number-13: 978-0-8493-9263-4 (Hardcover)

Library of Congress Cataloging-in-Publication Data

Chung, Yip-wah, 1950-
   Introduction to materials science and engineering / Yip-Wah Chung.
      p. cm.
   ISBN-13: 978-0-8493-9263-4 (alk. paper)
   ISBN-10: 0-8493-9263-2 (alk. paper)
   1. Materials science. I. Title.

TA403.C475 2007
620.1'1--dc22
                                           2006021809

Visit the Taylor & Francis Web site at
http://www.taylorandfrancis.com

and the CRC Press Web site at
http://www.crcpress.com

Preface

This book is based on lecture notes that I developed for an introductory course on materials science and engineering (MSE) at Northwestern University. It is suitable as a text for a one-semester course for science and engineering undergraduate students. To get the most out of the book, the reader should have at least one year of college physics and chemistry. All core topics normally covered in a semester of introductory MSE are included. Each chapter begins with some simple facts, a story, or an experiment, with the objective of showing the reader why the subject matter is relevant or important. Interesting facts and stories are introduced along the way, usually in the form of footnotes — partly to educate, but mostly to entertain. The reader may notice my passion for aviation. An airplane is one of the most amazing inventions in modern civilization; I always use it to illustrate what MSE is all about. Wherever appropriate, I also include examples of the inter-relationships among MSE, nanotechnology, and life sciences.

The book begins with three contemporary examples of why it is important to know something about MSE: airplanes, computer hard-disk drives, and prosthetic implants. Here I introduce the core concept of the structure–property relationship. This is followed by traditional topics, such as atomic/molecular bonding and crystal structures. On the subject of covalent bonding, I take the liberty of expanding the discussion on sp^3, sp^2, and sp bonding in carbon, primarily because of the importance of carbon-based materials. The section on x-ray diffraction includes applications in thin-film studies. Chapter 2 covers the subject of crystalline imperfections and diffusion. Students often feel that imperfections or defects are bad or undesirable. Teachers must convey that, in many cases, imperfections or defects are good because, without them, modern technologies, such as computers and planes, would not exist.

When I teach introductory MSE classes, I like to move to electrical properties as quickly as possible, primarily because I believe this is a comfort zone for students. Chapter 3 covers a lot of ground, from the energy band model of electronic conduction to semiconductors, transistors, lasers, and thin films. Chapter 4 begins with a short story of Paul MacCready, who successfully built a human-powered aircraft. Such a feat would not have been possible without the use of lightweight, high-strength materials and careful aerodynamics design. In addition to the usual topics in mechanical properties, I include a short section on statistics. Because mechanical properties tend to be sensitive to small compositional and microstructural changes, an engineer should be knowledgeable about the statistical characterization of these properties.

In my experience, students often find the subject of phase diagrams uninspiring. By using stories ranging from rockets to diamond synthesis to Chinese folk tales, I hope to bring some degree of interest and relevance to this subject. Chapter 5 on phase diagrams ends with topics on nano-crystalline materials and dental amalgams. The chapters on ceramics and polymers, Chapters 6 and 7, respectively, provide an overview, skipping topics related to processing because this is not an issue for most students who do not major in materials science. Materials science majors, however, will have subsequent courses on these topics that go into greater detail. The chapter on polymers also includes topics not normally covered in introductory texts — for example, a discussion of oxygen cross-linking and flame retardants for polymers.

Chapter 8 begins with a trick to clean silverware that is one of many examples of useful corrosion processes. Another useful corrosion process is one that leads to the generation of electrical power. I like the sections on energy analysis for electric vehicles because they show where advances are needed in terms of alternative energy sources. A brief summary of popular batteries is also included.

Chapter 9 concerns magnetic materials. Rather than repeating standard magnetism topics found in physics textbooks, this chapter explores several practical magnetism topics, such as power generation, data storage, magnetic resonance imaging, and surveying. One major difficulty in tackling engineering problems associated with magnetism is the multiple systems of units. I use this as an excuse to include a story about an aviation incident that resulted from mixing up units. Teachers should convey to their students in the strongest possible terms that getting units and arithmetic right is important and can mean the difference between life and death. The primary focus of Chapter 10 is the synthesis and selected properties of thin films. Despite the importance of thin films in modern technology, this topic is not included in other introductory MSE texts. Its inclusion, I hope, provides a more complete curriculum in the training of engineering undergraduates.

It has been fun as well as humbling to write this book. Morrie Fine, John Hilliard, and Lynn Johnson are sources of some of the stories in this book; they have my deepest admiration and respect. I am very grateful to Glenn Daehn, Narenda Dahotre, Michael Dugger, Rachel Goldman, Mark Hersam, Catherine Klapperich, Nikhil Koratkar, Hong Liang, Angus Rockett, Sudipta Seal, Susan Sinnot, Kathleen Stair, and Rodney Trice, who were kind enough to read and critique the manuscript; to my flight instructors, Patrick Kennedy, Bob Werneth, Joel DeJong, Paul Pieri, Rus Duczak, and David Stark, who unknowingly told me many aviation stories (some of which have found their way into this book); and to the many students taking my introductory MSE classes, who provided me with many useful comments. I wish to thank my daughters, Christina and Connie, for their unfailing trust and support. No words can truly express my gratitude, indebtedness to, and appreciation of my wife, Metty, for her love, encouragement, and support. Without her as a part of my life, I doubt whether this book would ever have been finished.

This book was written with the goal to inform, educate, and entertain. I hope to have accomplished at least a little bit of each. Have fun exploring.

Yip-Wah Chung
Wilmette, Illinois

The Author

Yip-Wah Chung obtained his Ph.D. in physics from the University of California at Berkeley. He joined Northwestern University in 1977 and is currently a professor of materials science and engineering there. Dr. Chung's research interests are in surface science, thin films, and tribology. He has been named a fellow of ASM International, American Vacuum Society, and the Society of Tribologists and Lubrication Engineers. His other awards include the Ralph A. Teetor Engineering Educator Award from the Society of Automotive Engineers; the Innovative Research Award and Best Paper Award from the American Society of Mechanical Engineers Tribology Division, Technical Achievement Award from the National Storage Industry Consortium (now Information Storage Industry Consortium), Bronze Bauhinia Star from the Hong Kong Special Administrative Government. He also served as an advisory professor at Fudan University in Shanghai.

Dr. Chung served two years as a program officer in surface engineering and materials design at the National Science Foundation. His most recent research activities are in low-friction surfaces, oxide coatings, and high-temperature friction phenomena. His favorite hobbies are photography and recreational flying. He holds several Federal Aviation Administration ratings, including commercial multiengine instrument, instrument ground instructor, and advanced ground instructor.

Contents

1 Introduction

1.1 WHAT IS MATERIALS SCIENCE AND ENGINEERING?

One way to appreciate the field of materials science and engineering is to look at commercial aviation. A Boeing 747 jumbo jet can transport 350 passengers and crew and their belongings from Chicago to Hong Kong (Figure 1.1) in less than 15 hours, consuming about 150 tons of fuel.* Aeronautical engineers design the Boeing 747 to be aerodynamically efficient. Electrical engineers design and build sophisticated avionics so that the 747 can fly from takeoff to touchdown. Who are the people designing and developing materials for the powerful engines, streamlined and pressurized fuselage, and flexible yet strong wings and control surfaces? You guessed right: materials scientists and engineers. In fact, about 150,000 lb of lightweight, high-strength aluminum alloys are used in the 747. If major structures of the 747 were made of steel instead of aluminum, the aircraft would have to eliminate all passengers and freight and still be over the maximum takeoff weight — clearly not a commercially viable proposition.

Think about computers. What is the most valuable component of a computer? It is the hard-disk drive, which stores important programs and data (Figure 1.2). In 1990, the cost of 1 Mbyte of data storage in a hard drive was about $1. In mid-2005, the same amount of data storage cost only $0.001, a 1000-fold reduction. In large part, this was made possible by major materials developments (improved materials for read–write heads, finer grained magnetic media, thinner and more wear-resistant protective coatings, etc.) that allow the read–write head to fly within a few nanometers of the disk surface. In this way, smaller bits can be written without catastrophic failure.**

We are now using various types of prosthetic devices in the human body. One example is artificial knee joints (Figure 1.3). Not only must the materials introduced into the human body be biocompatible, but they must also withstand the stress of daily use in a corrosive environment, without failure. The generation of large wear particles must be minimized because they may induce undesirable immune response.***

The search for the best materials in specific applications can be done by random trial and error, educated guesses or by more methodical approaches. In materials engineering, we must first define the set of properties needed to fulfill the performance requirements. For example, if we design a device with an operating temperature of 800°C, one obvious property of the material used for this device is that its melting point must be higher than 800°C. If the device needs to withstand certain stress levels at this temperature, then the material chosen must have the required strength

---

* The maximum takeoff weight for the B747-400 is 875,000 lb. The empty weight, depending on configuration, is 358,000 lb. Fuel (380,000 lb) is more than 40% of the weight of the fully loaded aircraft. That is why takeoff accidents can be dangerous. The maximum payload with full fuel is 137,000 lb.

** For most hard-disk drive systems in 1990, the read–write head was flying about 100 nm above the disk surface, providing an areal storage density of 100 Mbits/in.^{2}. In 2000, the flying height decreased to about 10 nm, and the areal storage density increased to about 10 Gbits/in.^{2}. When we scale the length of the read–write head (about 2 mm) to that of a B747 (about 70 m), this flying height corresponds to a jumbo jet flying at 0.35 mm above the ground. In spite of the close proximity between the head and the disk, the disk surface is protected from wear and catastrophic failure by advanced coatings and lubricants. It is projected that, by 2008, the flying height will decrease to 3 nm, and the areal storage density will increase to 1 Tbit/in.^{2}.

*** In prosthetic implants such as artificial hips and knees, it is estimated that each step taken by the patient may produce as many as 100,000 wear particles.

FIGURE 1.1 Boeing 747 on final approach to the Hong Kong Kai Tak Airport. (Reprinted with permission from Alexander Kueh, www.airliners.net.)

FIGURE 1.2 A hard-disk drive in mid-2005 stores about 80 Gbits/in.^{2}. (Reprinted with permission from Western Digital Corporation.)

at the operating temperature, with an appropriate safety margin. The next step is to examine the range of material composition and structure needed to provide the set of targeted properties. Finally, the materials engineer has to synthesize or fabricate the material with the required composition and structure (processing or synthesis). Figure 1.4 summarizes this process.

In general, the choice of materials and process is not unique. Compromises must be made as convenience, environmental concerns, availability, cost, and performance are balanced. The materials scientist, on the other hand, studies materials behavior in the opposite direction. For a given material, the materials scientist may examine how different methods of synthesis affect its structure. For example, amorphous carbon films can be deposited by directing low-energy carbon ions onto a substrate surface. The materials scientist may be interested in examining how the nearest-neighbor structure of these carbon films depends on the energy of the incident carbon ions. In fact, the carbon

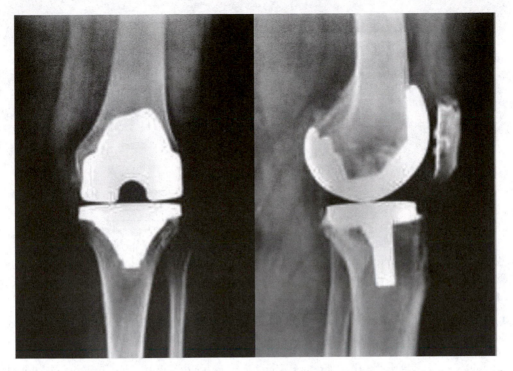

FIGURE 1.3 Front and side view of a knee after total knee replacement. (Reprinted with permission from Joint Replacement Institute, Los Angeles.)

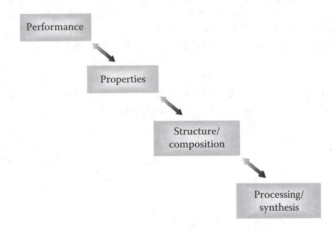

FIGURE 1.4 Materials science and engineering.

sp³ (diamond-like) fraction in these films attains a maximum at carbon ion incident energies ~ 100 eV (Figure 1.5). At this point, the materials scientist may wish to explore how this local structure affects the mechanical and electronic properties of carbon films and how such properties relate to performance in specific applications. This is the heart of materials science and engineering.

1.2 FUNDAMENTAL PRINCIPLES

Three fundamental principles govern structure and properties of materials. The first principle is: *Properties of a material depend on its structure*. For example, graphite and diamond are pure carbon, yet they have totally different properties. Graphite is soft and slippery and has a dark color

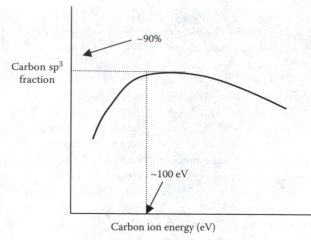

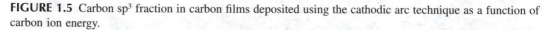

FIGURE 1.5 Carbon sp³ fraction in carbon films deposited using the cathodic arc technique as a function of carbon ion energy.

and good electrical conductivity, while diamond is hard, an electrical insulator (electrical resistivity greater than 10^{16} ohm-cm), and transparent from 225 nm to far infrared. These differences are due to the different atomic arrangements: Carbon atoms in graphite are organized in the form of hexagonal sheets, whereas carbon atoms in diamond are bonded to one another in fourfold tetrahedral arrangements.

Another example is the dependence of electrical resistivity on the concentration of gold in a copper–gold alloy. When small concentrations of gold are added into otherwise pure copper, gold atoms go into random lattice sites. This random arrangement of gold atoms disrupts the periodicity of an otherwise perfect lattice, resulting in an increase in the electrical resistivity. With further gold addition, the electrical resistivity begins to decrease and attains a minimum value when the gold concentration reaches 25 atomic percent. Further Au addition results in another resistivity minimum at 50 atomic percent (Figure 1.6). Both minima are associated with the formation of ordered Cu–Au alloys in which copper and gold atoms are sitting in well-defined lattice positions.

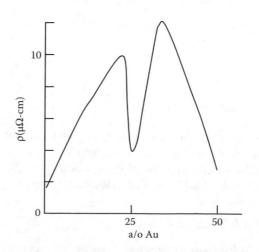

FIGURE 1.6 Room-temperature electrical resistivity versus gold concentration in a copper–gold alloy. (Adapted from Johansson, C. H. and J. O. Linde. 1936. *Annals of Physics* 25:1.)

The structure–property relationship discussed here refers to structure on the atomic scale. In many engineering materials, structure occurs at several levels beyond atomic. For example, a typical aluminum alloy used in a Boeing 747 is polycrystalline — that is, it consists of single crystal grains separated from one another by transition regions known as grain boundaries. In addition, precipitates are embedded inside these crystal grains. The average grain size, chemical composition, size, and concentration of these precipitates directly control mechanical properties of the alloy. In advanced solid-state lasers, multiple semiconductor layers are stacked together so that light emission and amplification occur with high efficiency at specific wavelengths. The composition, thickness, and hierarchy of stacking these layers control the operation of such lasers.

How is the structure of a given material determined? This is addressed by the second fundamental principle: *A material acquires a certain equilibrium structure in order to attain the lowest total free energy for the system.* We will give a precise definition of the term "free energy" later. For the purpose of the present discussion, we will simply equate that to the internal energy of the system. Let us use carbon as an example. The electronic configuration of a carbon atom can be written as: $(1s)^2 (2s)^2 (2p_x 2p_y)$. In this configuration, only electrons in the $2p_x$ and $2p_y$ orbitals are unpaired and available for chemical bonding. This should have resulted in a valence of two for carbon. If this were true, each carbon atom would only be able to bond with two other carbon atoms, and we would have quite a different universe.

Somehow, the system of carbon atoms is smart.* The 2s and 2p orbitals rearrange themselves to form four new states of the same energy. This is known as sp^3 hybridization. The hybridization is due to the 2s electrons being promoted to the $2p_z$ orbital. Of course, it costs energy to do so. However, this initial investment in energy is more than paid back by each carbon atom having a valence of four. Each carbon atom can now bond with four other carbon atoms, thus reducing the energy of the system. The carbon–carbon bond energy is about 4 eV. These sp^3-bonded carbon atoms form the basic tetrahedral structure of diamond.

The preceding principle only addresses the equilibrium structure. In many desirable and undesirable situations, the system is not at equilibrium. In that case, the structure will be governed by the kinetics of processes involved. For the polycrystalline aluminum alloys discussed earlier, the size of the precipitates controls mechanical properties. If allowed to proceed to equilibrium, these precipitates will grow in size (coarsen) from nanometers to microns, resulting in marked changes of mechanical properties. This leads us to the third principle: *Properties depend directly on synthesis or processing techniques.* In the next two sections, we will discuss the formation of solid structures from the building blocks of atoms and molecules.

Question for discussion: What everyday examples can you cite to demonstrate that properties depend on processing techniques?

1.3 ATOMIC AND MOLECULAR BONDING

Generally, atoms and molecules exert long-range attraction on one another. We will discuss the nature of such attractive interactions in a moment. As we bring atoms closer than the equilibrium distance, repulsive interaction begins to dominate. There are two reasons. One is the Coulomb repulsion between electrons and between positive ion cores of the approaching atoms or molecules. Another source of the repulsive interaction has a quantum origin. By forcing atoms closer together, we attempt to confine electrons into a smaller volume. Quantum mechanics shows that such confinement results in a higher energy for the system, representing a repulsive force on approaching atoms or molecules.** The distance at which the attractive force equals the repulsive force is the

---

* In this part of the galaxy, we have life forms and intelligence based on carbon. Carbon-based intelligent beings (we) are trying to create intelligence out of silicon (computers) — the element below carbon in the periodic table. One wonders if there are silicon-based life forms in another part of the universe trying to create intelligence out of carbon.

** Recall that force $F = -dE/dR$, where E is the energy and R distance.

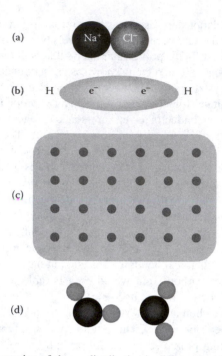

FIGURE 1.7 Schematic representation of charge distribution in different bonding types: (a) ionic, as in Na^+Cl^-; (b) covalent, as in the hydrogen molecule, where the electron cloud is located between two protons; (c) metallic, as in sodium (The filled circles represent sodium ion cores, and the gray background represents conduction electrons delocalized or spread out over the entire metal.); (d) dipole–dipole, as in water, which is charged positively on the hydrogen end and negatively on the oxygen end.

equilibrium separation for a given bond. We can broadly classify bonding into four types: ionic, covalent, metallic, and dipole, as discussed next (Figure 1.7).

1.3.1 IONIC BONDING

A good example of ionic bonding is sodium chloride (NaCl), a common ingredient of table salt. The electronic configuration of sodium is: $(1s)^2 (2s)^2 (2p)^6 (3s)^1$. The outermost 3s electron is loosely bound. The electronic configuration of chlorine is: $(1s)^2 (2s)^2 (2p)^6 (3s)^2 (3p)^5$. In this case, chlorine likes to acquire another electron to fill the 3p shell. When sodium and chlorine atoms are mixed together, the tendency is for sodium to lose electrons to chlorine, as indicated by the following equation:

$$Na + Cl \rightarrow Na^+ + Cl^-. \tag{1.1}$$

Considering just one pair of these ions, we can divide the interaction potential into two parts. The first part is Coulomb attraction between the ion pair, while the second part is repulsion as discussed previously. The repulsion is generally modeled as some inverse power of the distance between ions. Therefore, we can write the interaction potential U as:

$$U = -\frac{A}{R} + \frac{B}{R^n}, \tag{1.2}$$

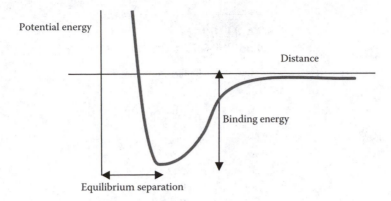

FIGURE 1.8 Potential energy as a function of distance between two atoms.

where

$$A = e^2/4\pi\varepsilon_o$$

e = electron charge

ε_o = permittivity of free space

B and n ($\approx$5–12) = empirical constants.*

Equation (1.2) is shown schematically in Figure 1.8. The depth of the potential energy well at the equilibrium distance is the cohesive or binding energy for the system.

In most ionic crystals, the attractive potential at the equilibrium distance is almost equal to the cohesive energy.** Therefore, larger ions tend to result in smaller cohesive energies. Table 1.1 lists

TABLE 1.1
Lattice Cohesive Energies of Selected Alkali Halides, Referenced to Free Ions

Compound	Cohesive Energy (kJ/mol)	Melting Point (°C)
LiCl	853	610
NaF	923	996
NaCl	786	801
NaBr	747	747
NaI	704	660
KCl	715	771
RbCl	689	715
CsCl	659	645

* A popular potential to represent interactions between neutral species is the Lennard–Jones 6–12 potential, which is written as $U = B[(\sigma/R)^{12} - (\sigma/R)^6]$. At short range, the potential increases rapidly with decreasing distance, representing a "hard wall." Atomic force microscopy, a popular surface imaging technique, relies on this hard-wall potential to obtain high vertical resolution in the contact-imaging mode.

** Have you ever thought about why water dissolves sodium chloride? Water has a dielectric constant of 80. This means that it reduces electrostatic attraction between two charged particles by a factor of 80, compared with free space. As a result, when sodium chloride is placed in water, the repulsive force between Na^+ and Cl^- takes over, resulting in dissolution of sodium chloride.

TABLE 1.2
Radii of Selected Ions

Ion	Radius (nm)
Li^+	0.060
Na^+	0.095
K^+	0.133
Rb^+	0.148
Cs^+	0.169
F^-	0.136
Cl^-	0.181
Br^-	0.195
I^-	0.216

the lattice cohesive energies of a series of halides and their melting temperatures. For example, single crystal sodium chloride has a cohesive energy of 786 kJ/mol and melting point of 801°C, while cesium chloride has a cohesive energy of 659 kJ/mol and melting point of 645°C. The cesium ion radius (0.169 nm) is significantly larger than the sodium ionic radius (0.095 nm). Likewise, single crystal sodium iodide has a cohesive energy of 704 kJ/mol and melting point of 660°C. In this case, the iodide ion radius (0.216 nm) is greater than the chloride ion radius (0.181 nm), as seen from Table 1.2.

EXAMPLE

Consider the Lennard–Jones 6–12 potential — that is, the interaction between two atoms $U(R)$ is given by:

$$U(R) = B\left[\left(\frac{\sigma}{R}\right)^{12} - \left(\frac{\sigma}{R}\right)^{6}\right].$$

Express the equilibrium distance and the binding energy of the system in terms of B and σ.

Solution

The system is at equilibrium when the interaction potential is a minimum — that is, $dU/dR = 0$. This gives:

$$12\left(\frac{\sigma}{R}\right)^{12}\frac{1}{R} = 6\left(\frac{\sigma}{R}\right)^{6}\frac{1}{R},$$

which reveals the equilibrium distance to be $2^{1/6}\sigma$ or 1.122σ. Substituting this equilibrium value of R into the original Lennard–Jones potential, we have the equilibrium binding energy of the system:

$$B\left[\left(\frac{\sigma}{2^{1/6}\sigma}\right)^{12} - \left(\frac{\sigma}{2^{1/6}\sigma}\right)^{6}\right] = \frac{1}{4}B - \frac{1}{2}B$$

$$= -\frac{1}{4}B.$$

1.3.2 COVALENT BONDING

One example of covalent bonding is the hydrogen molecule. Consider bringing two hydrogen atoms together, each with one electron in the 1s orbital. When the two atoms are close enough, the two electrons begin to interact or "couple." Such coupling affects energy levels of the combined system; instead of having two separate 1s orbitals, the two orbitals, labeled as σ (bonding) and σ* (anti-bonding) in Figure 1.9, are obtained. The energy position of these two levels depends on separation between the two atoms.

At some optimum distance, the σ level attains the lowest energy. Since this orbital can accommodate two electrons with opposite spins, lower total energy for the combined system can be obtained by causing both 1s electrons to fill this σ orbital. By the second fundamental principle stated in the preceding section, this reduction in total energy for the system drives the formation of the hydrogen molecule. The binding is generally described as due to the sharing of electrons between two hydrogen atoms and is known as covalent bonding.

To visualize covalent bonding in the hydrogen molecule, we can imagine an electron cloud concentrating mainly between the two hydrogen atoms. Distribution of the electron cloud is cylindrically symmetric about the bonding axis. There is very little electron concentration more than a few tenths of a nanometer outside the hydrogen molecule.

It should be noted that when dissimilar atoms are involved in covalent bonding, charge transfer may occur. In this case, the binding is due to a mixture of covalent and ionic bonding. The extent of charge transfer depends on the electronegativity difference between the two atoms. Electronegativity is a measure of the affinity of an atom for electrons. Using the Pauling electronegativity scale, we can calculate the percent ionic character of a bond between atoms A and B by:

$$\left(1 - \exp\left[-\frac{(x_A - x_B)^2}{4}\right]\right) \times 100 , \qquad (1.3)$$

where the x's represent the Pauling electronegativity of the respective elements.

Generally, electronegativity increases moving across the row of the periodic table from left to right and decreases going down the column. Among all naturally occurring elements, the most electronegative elements are fluorine (3.98), oxygen (3.44), chlorine (3.16), and nitrogen (3.04). The most electropositive elements are francium (0.7), cesium (0.79), rubidium (0.82), and potassium (0.82).

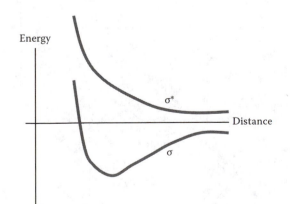

FIGURE 1.9 Formation of the hydrogen molecule.

sp³, sp², and sp Hybridization

As discussed in the previous section, the 2s and 2p orbitals of carbon rearrange themselves to form four new states of the same energy. This is known as sp³ hybridization. The geometric orientation of these four orbitals is shown in Figure 1.10. These are sometimes known as σ orbitals because the charge distribution is cylindrically symmetric about the bond axis.

In sp³ hybridization, we have four orbitals, s-p_x-p_y-p_z, mixed together. There is another way to mix these orbitals, represented symbolically as s-p_x-p_y + p_z. The first three are known as sp² hybridized orbitals. The sp² orbitals are planar, with bond angles of 120°. The p_z orbital is perpendicular to the plane defined by the sp² orbitals. In this configuration, two carbon atoms can form the double bond, one due to the overlap of sp² orbitals and one due to p_z. The p_z orbital overlap produces charge density above and below the plane defined by the sp² orbitals. This is sometimes known as π bonding (Figure 1.11).

There is yet another way to redistribute the four carbon orbitals, which can be symbolically represented by s-p_x + p_y + p_z. The resulting sp hybridized orbitals are shaped like an asymmetric dumbbell, as shown in Figure 1.12(a). With these rehybridized orbitals, carbon atoms form the triple bond, one from the overlap of sp orbitals, one from p_y, and one from p_z. Since the orientation of y- and z-axes is arbitrary, the resulting charge distribution due to the two p orbitals is a cylindrical sheet of electron charge around the bond axis, as shown in Figure 1.12(b).

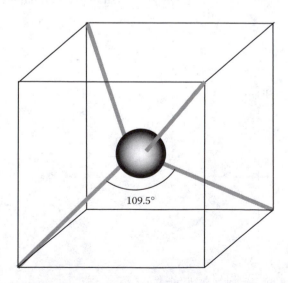

109.5°

FIGURE 1.10 Geometric orientation of sp³ orbitals, with the carbon atom sitting at the center of a cube.

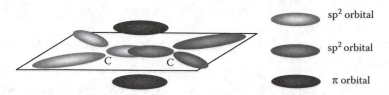

FIGURE 1.11 Charge (electron cloud) distribution in a carbon–carbon double bond.

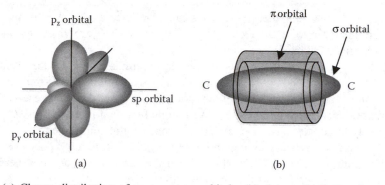

FIGURE 1.12 (a) Charge distribution of sp + p_y + p_z orbitals; (b) charge distribution in a carbon–carbon triple bond.

1.3.3 METALLIC BONDING

When we bring two or more metal atoms together, splitting of electron energy levels occurs similar to that observed in covalent bonding. For example, when we bring N sodium atoms together, we have N 3s valence electrons and N closely spaced 3s levels forming an energy band, as shown in Figure 1.13. Since each level can hold two electrons of opposite spin, the band is only half full, resulting in lower total energy for the system. This drives the formation of metallic bonds.

The major difference between covalent and metallic bonding is that these valence electrons are shared by all metal atoms and not bound to any specific atoms — that is, the valence electrons are

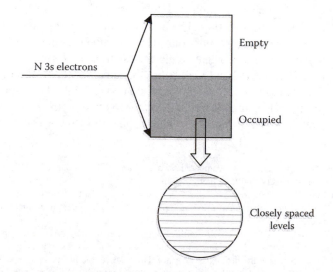

FIGURE 1.13 Formation of an energy band in a metal.

TABLE 1.3
Variation of Cohesive Energy and Melting Point of the Fourth-Row Elements of the Periodic Table

	K	Ca	Sc	Ti	V	Cr	Mn	Fe	Co	Ni	Cu	Zn
Cohesive energy (kJ/mol)	89	178	378	471	515	397	281	415	426	431	338	131
Melting point (°C)	64	838	1539	1668	1900	1875	1245	1536	1495	1453	1083	420

delocalized. These delocalized valence electrons can be viewed as the glue for the entire system, free to move around. Due to this delocalized property of the valence electrons, metals are good electrical and thermal conductors.

Take the glue analogy further: More glue results in stronger binding. For example, calcium, whose electronic configuration is [Ar] $(4s)^2$, has a cohesive energy of 178 kJ/mol and melting point of 838°C. Titanium has an electronic configuration of [Ar] $(3d)^2(4s)^2$. These two unpaired 3d electrons increase the cohesive energy to 471 kJ/mol and melting point to 1668°C.* When the 3d shell is completely filled with electrons, the 3d electrons cannot contribute to the cohesive energy. In this case, Zn, with an electronic configuration of [Ar] $(3d)^{10}(4s)^2$, has a cohesive energy of only 131 kJ/mol and melting point of 420°C. Table 1.3 lists the variation of cohesive energy and melting point for the fourth row elements of the periodic table.

1.3.4 DIPOLE BONDING

For neutral atoms and molecules with filled shells, the net charge is zero, and there are no unpaired valence electrons available for sharing. Therefore, there is no opportunity for ionic, covalent, and metallic bonding. For example, all inert gases have completely filled shells. The binding between inert gas atoms is quite weak; helium** has a boiling point at 1 atm of 4.2 K, argon 87.4 K, and xenon 166 K. The attractive interaction between these inert gas atoms (or closed shell atoms and molecules in general) is due to the formation of induced dipoles. Consider that at some instant, the electron cloud of a given inert gas atom is slightly distorted due to random fluctuations: There are more electrons in one part of the atom than in another. As a result, a dipole is formed. This dipole produces an electric field that acts on neighboring atoms, inducing them to form dipoles as well. The resulting dipole-induced dipole attraction is known as van der Waals interaction. This interaction potential decays rapidly with separation R $(\propto 1/R^6)$.***

Some molecules are permanent dipoles. The resulting dipole–dipole interaction contributes to cohesion between these molecules. An important type of dipole bonding is hydrogen bonding. When hydrogen is bound to an electronegative element such as oxygen or nitrogen, the resulting bond is polarized; the OH and NH bond are slightly negative on the oxygen and nitrogen ends, respectively, while the hydrogen end is slightly positive (Figure 1.14). Figure 1.15 shows two molecules: thymine on the left and adenine on the right. Note how these two molecules bond to each other due to hydrogen bonding and the unique geometry of these two molecules. These are

---

* These two extra d electrons in titanium (as compared with calcium) make it one of the best elements for aviation applications: high strength, low density, and high melting point.
** Do you know that helium was first discovered on the Sun, rather than on Earth? Scientists studied light emission from the Sun and discovered spectral lines that could only be explained by the presence of an unknown element. The word "helium" is derived from *helios* (the Sun).
*** A modern hard-disk drive runs with the read–write head flying at a small height above the spinning disk surface. In designing the shape of the read–write head, the concern is about aerodynamic forces providing the required stable flying height, ignoring van der Waals forces at the head–disk interface. However, as the head comes to within a few nanometers, the attractive component of this force becomes dominant, making flying unstable. This is a major problem in pushing disk drives to ever increasing storage densities.

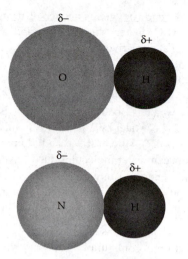

FIGURE 1.14 Charge distribution in OH and NH bonds.

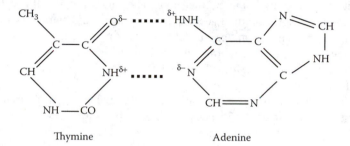

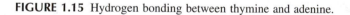

Thymine   Adenine

FIGURE 1.15 Hydrogen bonding between thymine and adenine.

two of the DNA bases (the other two are cytosine and guanine). This dipole–dipole bonding (hydrogen bonding) results in the formation of the double-helix structure of DNA.

EXAMPLE

Calculate the degree of ionic character of the OH bond, given that the electronegativity of oxygen is 3.44 and that of hydrogen is 2.2.

Solution

Since $x(O) = 3.44$ and $x(H) = 2.2$, $x(O) - x(H) = 1.24$. From Equation (1.3), the percent ionic character = $100 [1 - \exp(-1.24^2/4)] = 32\%$.

It could even be argued that, without hydrogen bonding, life would not have been possible for the formation of DNA and related molecules, as well as for the presence of liquid water on Earth. Consider the boiling points of hydrogenated compounds for group VI elements:

Compound	Boiling Point (°C)
H_2S	−60.7
H_2Se	−41.5
H_2Te	−2.0

Extrapolation of the indicated trend (oxygen is above sulfur in group VI of the periodic table) suggests that the boiling point of H_2O (water) should be around –80 to –100°C, as opposed to the actual value of 100°C. The additional attractive interaction due to hydrogen bonding between water molecules raises the boiling point by almost 200°C. Life would certainly be quite different if the boiling point of water were –100°C.

Another consequence of hydrogen bonding in water is the high specific heat of water. At room temperature, the specific heat of water is about 1 cal/g-K; that is, it takes 1 cal (4.18 J) of heat energy to raise the temperature of 1 g of water by 1 K. Compare this number with the specific heat of common materials such as copper (0.2), silicon (0.1), and aluminum (0.3). It takes more thermal energy to overcome attraction between water molecules due to hydrogen bonding. Therefore, water acts as an effective reservoir of heat energy. This explains why (1) lakes and oceans have moderating effects on local temperatures; and (2) air with high relative humidity tends to experience much smaller diurnal temperature variation than air with low relative humidity (e.g., in deserts).

1.4 CRYSTAL STRUCTURES

Almost all materials form crystals with regular atomic arrangements when they solidify. A crystal is made up of unit cells repeating in three dimensions. The smallest unit cell is called the primitive unit cell. Crystal structures are classified according to the shape and symmetry of their respective primitive unit cells. Each primitive unit cell is defined by six parameters: three unit vectors (lattice parameters) that define the side lengths of the unit cell and three angles between three pairs of unit vectors (Figure 1.16). Based on the values of relative unit vector lengths and angles, we can identify seven crystal systems, as shown in Figure 1.16.

In the following sections, we give examples and selected properties for three of the most common crystal structures (Figure 1.17): body-centered cubic (BCC), face-centered cubic (FCC), and hexagonal close-packed (HCP) structures.

1.4.1 BODY-CENTERED CUBIC (BCC)

An example of BCC is iron at room temperature or titanium above 882°C. This unit cell has one atom at the cube center and eight atoms at the corners of the cube. Since each cube corner is shared with eight other cubes, the BCC unit cell contains two atoms. Using the atom at the cube center as a reference, we can see that the coordination number (number of nearest neighbors) is eight. For cube side length a and atom radius R, assuming that the atom at the body center is in contact with its eight nearest neighbors, we can show that $a\sqrt{3} = 4R$ (see following example). From this information, we can calculate the ratio of volume occupied by the two atoms in the unit cell (assumed spherical) to the volume of the unit cell. This ratio is known as the packing factor. For the BCC unit cell, the packing factor is equal to $\pi\sqrt{3}/8$, about 68%.

EXAMPLE

For the BCC unit cell, determine the relationship between the cube side length a and the atomic radius R. Calculate the packing factor.

Solution

Consider the arrangement of atoms along the body diagonal as shown (Figure 1.18). The length of the body diagonal is equal to $a\sqrt{3}$, and there are four atomic radii along the diagonal. Therefore, we have $a\sqrt{3} = 4R$, or $R = a\sqrt{3}/4$.

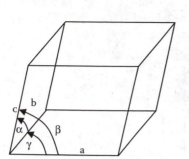

FIGURE 1.16 Seven crystal systems.

Crystal system	Unit vectors	Angles
Cubic	a = b = c	$\alpha = \beta = \gamma = 90°$
Tetragonal	a = b ≠ c	$\alpha = \beta = \gamma = 90°$
Hexagonal	a = b ≠ c	$\alpha = \beta = 90°, \gamma = 120°$
Orthorhombic	a ≠ b ≠ c	$\alpha = \beta = \gamma = 90°$
Rhombohedral	a = b = c	$\alpha = \beta = \gamma \neq 90°$
Monoclinic	a ≠ b ≠ c	$\alpha = \gamma = 90° \neq \beta$
Triclinic	a ≠ b ≠ c	$\alpha \neq \beta \neq \gamma$

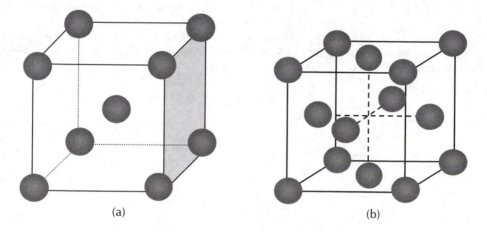

(a)

(b)

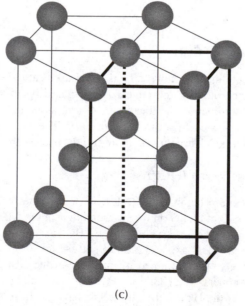

(c)

FIGURE 1.17 Atomic arrangements for three crystal structures: (a) body-centered cubic; (b) face-centered cubic; and (c) hexagonal close packed.

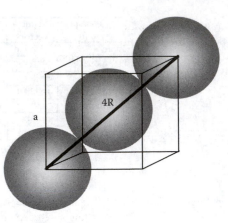

FIGURE 1.18 Atomic arrangements along the body diagonal of a body-centered cubic unit cell.

The total volume of the unit cell is a^3. As discussed earlier, there are two atoms per unit cell. The volume of these two atoms is equal to $(8/3)\,\pi R^3$. The packing factor (PF) is given by the ratio of volume of atoms to volume of unit cell:

$$PF = \frac{(8/3)\pi R^3}{a^3}$$

$$= \frac{8}{3}\pi\left(\frac{\sqrt{3}}{4}\right)^3$$

$$= \frac{\pi\sqrt{3}}{8}$$

$$= 0.68.$$

1.4.2 FACE-CENTERED CUBIC (FCC)*

An example of FCC is aluminum. This unit cell has eight atoms at the cube corners and six atoms at the cube faces. Since each atom at the cube face is shared by two cubes, the FCC unit cell contains four atoms. The coordination number is 12. Again, assuming close packing with nearest neighbors and spherical atoms, we can show that $a\sqrt{2} = 4R$. The packing factor is equal to $\pi\sqrt{2}/6$, or about 74%. Therefore, the FCC structure is more tightly packed than the BCC structure.

1.4.3 HEXAGONAL CLOSE PACKED (HCP)

An example of HCP is titanium below 882°C. The HCP structure can be obtained by stacking close-packed hexagonal planes of atoms on top of each other in the ABAB stacking sequence; that is, the stacking arrangement repeats every two layers. By convention, the unit vector along this packing direction is known as the c-axis, and the a- and b-axes are perpendicular to it. The FCC structure, on the other hand, can be obtained by stacking in the ABCABC sequence. Therefore, the nearest neighbor environments for HCP and FCC are similar. The coordination number (12) and packing factor (74%) for the HCP unit cell are the same as those for FCC.

---

* The diamond structure can be considered to be two interpenetrating FCC unit cells, with the second FCC unit cell displaced by one quarter of the way along the body diagonal.

For perfectly spherical atoms, the c/a ratio in the HCP crystal should be equal to $2\sqrt{2}/\sqrt{3} = 1.633$. This can be compared with the c/a ratio for Mg (1.623), Zn (1.856), and Ti (1.587). This comparison indicates that Mg atoms are close to spherical, Zn atoms slightly elongated, and Ti atoms slightly compressed along the c-axis.

EXAMPLE

Determine the c/a ratio of an ideal HCP crystal.

Solution

Consider the four-atom cluster shown in Figure 1.19. The lower three atoms form part of the A plane (as in the ABAB repeat sequence discussed in the preceding paragraph on the HCP structure) and the upper atom is part of the B plane. The lower three atoms form an equilateral triangle with side length a (= 2R; R is the atom radius). The upper atom sits above the center of this triangle as shown in Figure 1.19(b).

With this information, we can construct the triangle formed by the topmost atom and two atoms on the A plane, with side length as shown. Applying the Pythagorean theorem, we have:

$$(2R)^2 = \left(\frac{c}{2}\right)^2 + \left(\frac{2}{3}\sqrt{3}R\right)^2.$$

Solving, we have:

$$c = 4\sqrt{\frac{2}{3}}R.$$

Therefore,

$$\frac{c}{a} = \frac{c}{2R} = 2\sqrt{\frac{2}{3}} = 1.633.$$

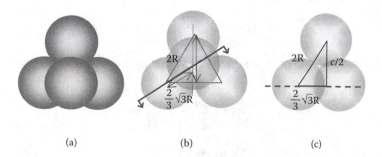

(a)   (b)   (c)

FIGURE 1.19 (a) Four-atom cluster for an ideal hexagonal close-packed unit cell; (b) dimension of the equilateral triangle formed by the lower three atoms (top view); (c) view of (b) taken from a vertical section along the solid line.

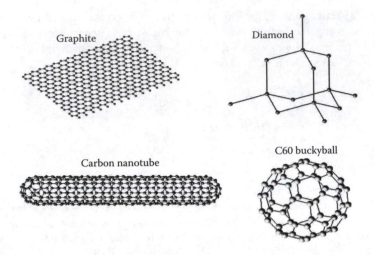

FIGURE 1.20 Different forms of carbon. (Reprinted with permission from Prof. Dong Qian, University of Cincinnati.)

1.5 POLYMORPHISM

A given material may crystallize in more than one structure with the same composition. These different possible structures are known as polymorphs (poly = many; morph = form) or allotropes. Generally, we use the term polymorphs to refer to compounds and allotropes to refer to elements. Carbon can exist as graphite, diamond, and various forms of nanotubes and buckyballs (Figure 1.20). Iron has a BCC structure at room temperature and transforms to FCC above 912°C. At room temperature, tin has a tetragonal structure (β-Sn, a metal) and transforms into a diamond structure below 13°C (α-Sn, a semiconductor).* We will discuss this more in the chapter on phase diagrams.

EXAMPLE

What is the percentage volume contraction when an element undergoes the BCC to FCC structural transformation? Assume that there is no change in the atomic size during this transformation.

Solution

Let us first look at the BCC unit cell, with cube side length a. As discussed earlier, there is a relationship between a and R, the atomic radius:

$$a\sqrt{3} = 4R.$$

Since there are two atoms per BCC unit cell, we can calculate the volume occupied by each atom:

$$\frac{a^3}{2} = \frac{1}{2}\left(\frac{4R}{\sqrt{3}}\right)^3 = \frac{32}{3\sqrt{3}}R^3$$

$$= 6.1584R^3.$$

---

* The transformation from tetragonal to cubic tin involves a significant volume change, resulting in crumbling. In the past, this was called tin disease. The story of Napoleon's troops' (tin) armor disintegrating while invading Russia in the depth of winter is very likely due to this tin disease.

We can repeat the same calculation for the FCC unit cell, remembering that there are four atoms per FCC unit cell. The volume occupied by each atom in that cell is then given by:

$$\frac{a^3}{4} = \frac{8}{\sqrt{2}} R^3$$

$$= 5.6569 R^3,$$

which represents a contraction. The percentage contraction in going from BCC to FCC is equal to:

$$\frac{6.1584 - 5.6569}{6.1584} \times 100 \approx 8\%.$$

1.6 LABELING DIRECTIONS AND PLANES

To label directions in a unit cell of a given crystal structure, first draw a parallel ray passing through the origin as shown in Figure 1.21(a). The direction is labeled by the Cartesian coordinates of a point on that ray reduced to the set of smallest integers. The direction is written as [XYZ]. Note the use of square brackets. A group of equivalent directions is written within angled brackets <...>. For example, for cubic unit cells, [100], [010], etc. are equivalent and can be represented by <100>. The angle between two directions [u,v,w] and [u′,v′,w′] is given by

$$\cos^{-1} \frac{uu' + vv' + ww'}{\sqrt{u^2 + v^2 + w^2} \sqrt{u'^2 + v'^2 + w'^2}}. \tag{1.4}$$

EXAMPLE

Draw the [211] direction for a cubic unit cell.

Solution

Locate the point (2,1,1) as shown in Figure 1.21(b). The direction is the line drawn from the origin to this point.

EXAMPLE

Consider the line from the point A(1,2,0) to B(3,1,1). What is this direction?

Solution

We need to translate the line so that the first point is at the origin. We can accomplish this by a subtraction procedure:

x-coordinate for point B $\rightarrow$ 3 − 1 = 2
y-coordinate for point B $\rightarrow$ 1 − 2 = −1
z-coordinate for point B $\rightarrow$ 1 − 0 = 1

The new coordinate for point B is therefore (2,−1,1), as shown in Figure 1.21(c). The direction is written as [2$\bar{1}$1]. Note the bar notation for negative values.

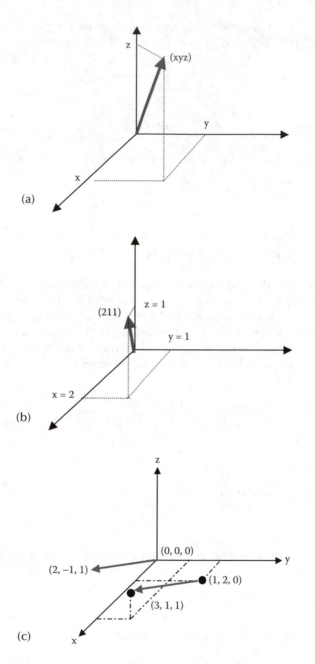

FIGURE 1.21 (a) Labeling direction [XYZ]; (b) labeling direction [211]; (c) determining the direction of line from (1,2,0) to (3,1,1).

To label planes, we first let the plane intercept the three axes as shown in Figure 1.22(a). The intercept values are a, b, and c. If the plane is parallel to a certain axis, the intercept value is infinity. Then, take the reciprocal of each of these values, and reduce to a set of smallest integers. The plane is labeled as (hkl). The label is known as a Miller index. The family of equivalent planes is labeled as {hkl}. For example, in a cubic system, (100) plane is equivalent to (010) and (001). These planes can be represented by {100}.

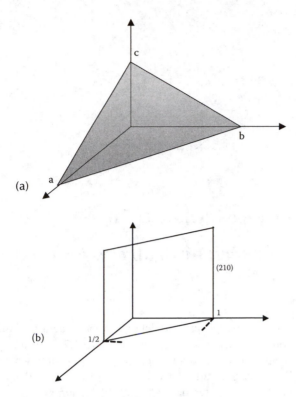

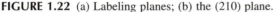

FIGURE 1.22 (a) Labeling planes; (b) the (210) plane.

EXAMPLE

Draw the (210) plane for the cubic unit cell.

Solution

The first step is to identify the intercepts on all three axes. The (210) plane intercepts the x-axis at 1/2 (reciprocal of 2) and the y-axis at 1. For the z-axis, the intercept is infinity (1/0). This means that the (210) is parallel to the z-axis. The result is shown in Figure 1.22(b).

For cubic systems, the spacing d_{hkl} between {hkl} planes is given by:

$$d_{hkl} = \frac{a}{\sqrt{h^2 + k^2 + l^2}},$$
(1.5)

where a is the lattice parameter or side length of the unit cell.

1.6.1 HEXAGONAL CRYSTALS

It is customary to use the four-axis coordinate system to label directions and planes in hexagonal crystals (Figure 1.23). The first three axes (a_1, a_2, and a_3) lie on the basal (hexagonal) plane at 120° to one another. The fourth axis is along the z-axis perpendicular to the basal plane. For the direction shown in Figure 1.18, it is labeled as [u v w t], in which u, v, and t are as shown, and $w = -(u + v)$.

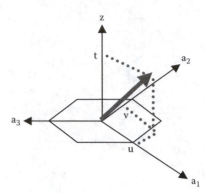

FIGURE 1.23 Four-axis coordinate system for a hexagonal structure.

1.7 DETERMINATION OF STRUCTURE AND COMPOSITION USING X-RAYS

1.7.1 X-Ray Diffraction

X-ray diffraction is the primary technique used to determine the structure of materials.* Consider the diffraction of x-rays from a family of {hkl} planes (Figure 1.24). In this geometry, the path difference between x-rays scattered from two adjacent planes is $2d_{hkl}\sin\theta$, where d_{hkl} is the spacing between {hkl} planes. When this path difference is equal to an integral multiple of the x-ray wavelength λ, we have constructive interference (a bright diffraction beam):

$$2d_{hkl}\sin\theta = n\lambda , \qquad (1.6)$$

where n is an integer. Therefore, measurement of the diffraction angle gives d_{hkl} and hence the size of the unit cell. Because of symmetry, not all families of planes can give rise to bright diffraction spots. BCC crystals give constructive interference for those $\{h\ k\ l\}$ planes when $h + k + l$ is an even integer. For FCC crystals, h, k, and l must be all odd or even. The following lists the planes giving rise to bright diffraction spots:

BCC: {110}, {200}, {211}, {220}, {310}, {222}
FCC: {111}, {200}, {220}, {311}, {222}

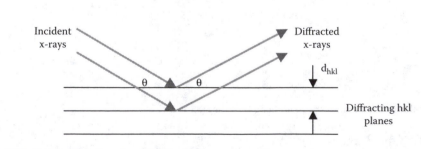

FIGURE 1.24 X-ray diffraction from {hkl} planes.

---

* Other tools are available for direct structural determination, such as neutron diffraction, electron diffraction, and scanning probe microscopy. Neutron diffraction can be performed only at nuclear reactor facilities. Electron diffraction is useful for thin samples because of limited electron penetration and must be performed in vacuum. Scanning probe microscopy is applicable to surfaces only.

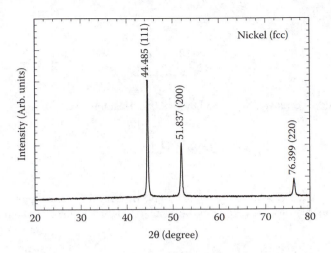

FIGURE 1.25 X-ray diffraction pattern from polycrystalline nickel.

While it is easy to obtain the interplanar spacing d_{hkl}, it is far more difficult to obtain the atomic structure from the x-ray diffraction pattern.* Figure 1.25 shows the x-ray diffraction obtained from polycrystalline nickel, which has an FCC structure.

EXAMPLE

Consider x-rays diffracting from the (111) planes of aluminum. The spacing between these planes is 0.233 nm. The wavelength of x-rays is 0.154 nm. Determine the diffraction angle θ for first-, second-, and third-order diffraction.

Solution

In this example, $d_{hkl} = 0.233$ nm, and $\lambda = 0.154$ nm. For first-order diffraction, $n = 1$. Using Equation (1.6), we can write:

$$\sin \theta = \frac{\lambda}{2d_{hkl}},$$

which gives $\theta = 19.3°$. Repeating the same calculation for second- and third-order diffraction, we obtain $\theta = 41.4$ and $82.5°$, respectively. Note that there is no fourth-order diffraction. When we calculate n $\lambda/2\ d_{hkl}$ for $n = 4$, the answer is >1. Since $\sin \theta$ cannot be >1, there is no solution. This means that there is no fourth-order diffraction.

EXAMPLE

How do we distinguish between BCC and FCC crystals using x-ray diffraction?

Solution

For BCC crystals, the first two planes visible by x-ray diffraction are (110) and (200). Using Equation (1.6), we can write:

$$\lambda = 2d_{110} \sin \theta_{110} = 2d_{200} \sin \theta_{200}.$$

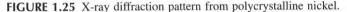

---

* It is beyond the scope of this book to go into the specifics of structure determination from x-ray diffraction. It suffices to say that the structure information is contained in the phase of the scattered x-rays, which is not directly obtainable from the diffraction intensity.

Therefore,

$$\frac{\sin\theta_{110}}{\sin\theta_{200}} = \frac{d_{200}}{d_{110}}.$$

For a unit cell side length a, $d_{200} = a/2$, and $d_{110} = a/\sqrt{2}$. Therefore, we have:

$$\frac{\sin\theta_{110}}{\sin\theta_{200}} = \frac{1}{\sqrt{2}} = 0.707.$$

For FCC crystals, the first two diffraction planes visible by x-ray diffractions are (111) and (200). It is left to the reader to show that the ratio of the diffraction positions for the first two diffraction spots is:

$$\frac{\sin\theta_{111}}{\sin\theta_{200}} = \frac{\sqrt{3}}{2} = 0.866.$$

This is readily distinguishable from the BCC case.

Equation (1.6) suggests an important application for x-ray diffraction. If a material is subjected to mechanical stress, the spacing between atomic planes will change. This implies that the diffraction angles will shift according to the stress or strain applied to the material. Taking the derivative of Equation (1.6), we have:

$$\frac{\Delta d_{hkl}}{d_{hkl}} = -(\cot\theta)\Delta\theta . \qquad (1.7)$$

X-ray diffraction therefore provides a direct method to determine the internal stress or strain of materials from the shifting of diffraction peaks. As we will learn later in the chapter on mechanical properties, internal flaws in a material can amplify stress to high levels, which may cause catastrophic failure if left undetected and not remedied in a timely manner. This technique is therefore widely used for nondestructive evaluation of mechanical components (e.g., fuselage and control surfaces in aircraft).

EXAMPLE

We are attempting to measure the strain exerted on a given sample by measuring the position of a given x-ray diffraction spot at an angle $\theta = 45°$. With a strain of 0.01%, calculate the shift in the angular position of that diffraction spot. What is the angular position shift for another diffraction spot at $\theta = 85°$? Based on this comparison, comment on the choice of diffraction spots to obtain the best sensitivity in using x-ray diffraction to measure strain.

Solution

Strain $= \Delta d/d$. According to Equation (1.7), we can calculate the angular shift $\Delta\theta$ to be:

$$\Delta\theta = -\frac{1}{\cot\theta}\frac{\Delta d}{d}.$$

In this example, $\Delta d/d = 0.01\% = 0.0001$, and cot $45° = 1$. Solving, we obtain $\Delta\theta$ to be 0.0001 rad, or $0.0057°$, or 20.6 arc seconds. The corresponding angular position shift at $\theta = 85°$ can be similarly calculated to be $0.065°$, or about 236 arc seconds. This shows that we should choose diffraction spots with large diffraction angles to maximize the sensitivity for strain measurements using x-rays.

1.7.2 OTHER APPLICATIONS OF X-RAY SCATTERING

X-rays can be used to study properties not only of bulk materials, but also of thin films and surfaces. At sufficiently high x-ray energies, the refractive index of any solid is slightly less than one. As is known from standard optics, below a certain critical angle, total external reflection results. The critical angle in radians (θ_o) as measured from the surface when this occurs is given by:

$$\theta_o = \lambda \sqrt{\frac{nr_e}{\pi}}, \tag{1.8a}$$

where λ = x-ray wavelength, n = total electron density of the solid, and r_e = classical radius of electron* (2.81×10^{-15} m).

Therefore, measurement of the critical angle directly gives the electron density. For a sample of known composition, the mass density can be computed directly from electron density. For elemental solids, Equation (1.8b) relates the mass density ρ (g/cm^3) to the critical angle in degrees (θ_o), assuming x-rays of wavelength 0.154 nm as obtained from a copper x-ray source**:

$$\rho = 23.82 \frac{A}{Z} \theta_o^2, \tag{1.8b}$$

where A is the atomic weight of the material and Z the atomic number.

Figure 1.26 shows the scattered x-ray intensity versus scattering angle obtained from a nitrogenated carbon film. The critical angle as estimated from the angular position at which the intensity is equal to half the maximum is equal to $0.22°$. The x-rays are derived from a copper x-ray source so that Equation (1.8b) is applicable. The mass density ρ is then given by:

$$23.82 \frac{12}{6} (0.22)^2 = 2.31 \text{ g/cm}^3.$$

Another application is to measure the thickness of thin films. Consider a thin film of thickness t deposited on a substrate; an x-ray beam, wavelength λ, is incident on the film at a small angle θ, as shown in Figure 1.27. Note that this configuration is identical to Figure 1.24, except that, in this case, we are considering the interference of x-rays scattered from the top and bottom of the

---

* The classical radius of an electron r_e is defined by the equation:

$$\frac{e^2}{4\pi\varepsilon_o r_e} = mc^2,$$

where e = electron charge, ε_o = permittivity of free space, m = electron mass, and c = speed of light. The left-hand side of the equation is the electrostatic energy due to a particle of radius r_e and charge e. The right-hand side represents the rest mass energy of the electron.

** This method is equivalent to weighing the material. In the total reflection geometry, x-rays typically penetrate less than 10 nm into the surface. For a lightweight material such as aluminum, a 10-nm slab weighs 2.7 μg/cm^2 — not a bad sensitivity just by bouncing around x-rays.

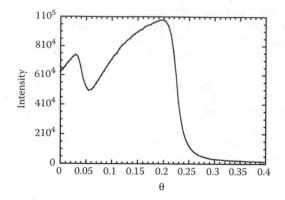

FIGURE 1.26 Low-angle scattered x-ray intensity versus angle obtained from a nitrogenated carbon film. Note the large increase in scattered intensity below 0.22°.

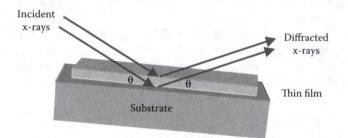

FIGURE 1.27 X-ray diffraction from a thin film.

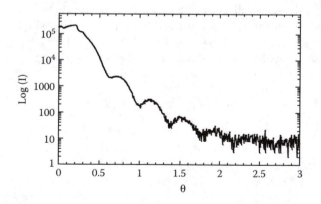

FIGURE 1.28 Low-angle x-ray diffraction from a nitrogenated carbon thin film deposited on a Si substrate, using 0.154-nm wavelength x-rays. Note the periodic maxima and minima.

film surface and the diffraction angle θ is smaller. The analogous equation for constructive interference is given by:

$$2t\theta = n\lambda .\tag{1.9}$$

Here, we equate sin θ to θ because the angles are typically small. Figure 1.28 shows the scattered x-ray intensity versus angle from a nitrogenated carbon thin film deposited on silicon, using x-rays with wavelength of 0.154 nm. Note two features from this plot: (1) multiple peaks beyond 0.3°, due to interference of x-rays scattered from the top and bottom surfaces of the film (from which we can obtain the film thickness through Equation 1.9); and (2) the rapid intensity decay with increasing θ (from which we can extract surface roughness information).

EXAMPLE

Determine the thickness of the nitrogenated carbon film shown in Figure 1.28.

Solution

From Figure 1.28, the average spacing θ between intensity maxima is estimated to be 0.42°, or 0.00733 rad. Using Equation (1.9), we have:

$$t = \frac{\lambda}{2\theta} = \frac{0.154}{2 \times 0.00733}$$

$$= 10.5.$$

Therefore, the film thickness is 10.5 nm.

Equation (1.6) indicates that there is a unique diffraction angle for a given set of diffraction planes. This is true only when we deal with an infinite crystal. With crystal grains of diameter D, the diffracted beam exhibits some amount of broadening $\Delta\theta$, given by:

$$\Delta\theta \approx \frac{\lambda}{D\cos\theta} , \tag{1.10}$$

where λ is the x-ray wavelength and θ the diffraction angle (as shown in Figure 1.24). This is known as the Debye–Scherrer formula and provides a powerful means to measure the average grain size of a given solid.

1.7.3 COMPOSITION DETERMINATION FROM EMISSION OF CHARACTERISTIC X-RAYS

When a material is bombarded by high-energy electrons or x-rays, characteristic x-ray emission results (i.e., the energy of emitted x-rays is characteristic of the elemental composition of the material). For example, consider the irradiation of an atom by electrons or x-rays with energy greater than the binding energy of K-shell (1s) electrons (Figure 1.29). There is a certain probability that the K-shell electron of the atom will be knocked out, leaving behind a vacancy. When this occurs, an L-shell electron falls down to fill the vacancy, releasing energy equal to the difference between these two energy levels.

The energy can be released in several forms. One form is x-ray emission. Since each element has a distinct set of electron energy levels, energies of emitted x-rays are characteristic of the element. Therefore, measuring energies of emitted x-rays provides a direct method of elemental identification. Coupled with the use of composition standards, this provides a powerful method for quantitative analysis of material composition, with typical sensitivity in the 0.1–1% range.*

When x-rays are used, the technique is known as x-ray fluorescence. When the electron beam from a scanning or transmission electron microscope is used to excite x-ray emission, the technique

---

* This technique is less sensitive towards light elements (up to neon), for two reasons. First, light elements do not emit x-rays with high efficiency. Second, x-rays emitted by light elements have lower energies and can be absorbed more easily by x-ray detector window materials.

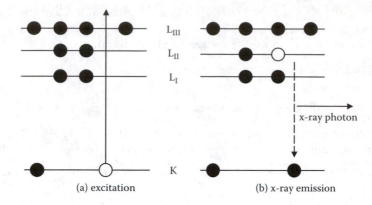

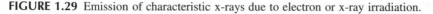

FIGURE 1.29 Emission of characteristic x-rays due to electron or x-ray irradiation.

is known as electron-beam microprobe or energy-dispersive x-ray analysis. In the latter case, the spatial resolution for bulk specimens is limited by the spreading of the electron beam and is typically on the order of a micron. Therefore, this technique allows composition analysis from small samples.

Question for discussion: This technique detects all elements except hydrogen and helium. Why are these two elements undetectable by the x-ray technique? (Hint: In order for x-rays to be emitted, it is necessary for one electron from a higher energy level to "fall" into a lower energy level.)

1.8 WHAT IS NEXT?

Studying the structure–property relationship is the central theme of materials science and engineering. We now have sophisticated tools such as x-ray diffraction to determine the atomic structure of materials.* Over the years, researchers have learned that materials generally prefer to form almost perfect crystal structures when they solidify. The operative word here is *almost* — imperfections exist in all real materials.** What are these imperfections? Why do imperfections exist? Is it bad to have an imperfect structure? Let us find out in the next chapter.

PROBLEMS

1. Given a positive and a negative ion, the attractive potential energy between these two ions is given by $-A/R$ and the repulsive potential energy by B/R^n, where R is the separation between the two ions. A, B, and n are constants. The total potential energy U is therefore equal to:

$$U = -\frac{A}{R} + \frac{B}{R^n}.$$

   a. When the system is at equilibrium, U is a minimum. Solve for the equilibrium separation between the two ions in terms of A, B, and n.
   b. Determine the binding energy for the ion pair at this equilibrium separation in terms of A, B, and n.

---

* Structures of the most complex biochemicals, such as proteins and DNA, are obtained by x-ray diffraction.
** Do you know that a single crystal cubic zirconia (fake diamond) that is more perfect than a natural diamond can be grown? Unfortunately (or fortunately, depending on your perspective), the value of a diamond is not entirely determined by crystalline perfection.

2. Make sketches to indicate the following directions in a cubic unit cell:
   a. [111]
   b. [200]
   c. [321]
3. Make sketches to indicate the following planes in a cubic unit cell:
   a. (111)
   b. (200)
   c. (321)
4. A line is drawn in a cubic unit cell from the point (1,1,0) to (3,1,1). What is this direction?
5. Show that the angle in the tetrahedral structure shown in Figure 1.10 is 109.5°.
6. Show that the packing factor for a face-centered cubic structure is 74%.
7. CsCl has a body-centered cubic structure. Consider the unit cell in which Cs^+ is at the center and Cl^- at the corners. Given that the radius for Cs^+ is 0.169 nm and that for Cl^- it is 0.181 nm, calculate the packing factor for this unit cell. Note: The packing factor referred to in the text for the BCC structure is for the case when all atoms in the unit cell are identical. Here, the two ions have different sizes.
8. X-rays of wavelength λ are diffracted by an iron sample. The first-order diffraction angle θ is 22.35° for the (110) planes.
   a. What is the wavelength of the x-rays used? The lattice constant of iron is 0.287 nm.
   b. For the same sample, same planes, and same x-rays, what is the angle for third-order diffraction? Be careful.
9. N, P, As, and Sb all belong to group V in the periodic table (in increasing atomic number). The boiling and melting point data for the hydrides are given in the following table:

Compound	Boiling Point (°C)	Melting Point (°C)
NH_3	−33.3	−???
PH_3	−87.75	−133
AsH_3	−62.5	−116
SbH_3	−17.0	−88

   a. Explain why NH_3 has a higher boiling point than expected from the trend. Note that nitrogen is the second most electronegative element (after fluorine).
   b. Estimate the melting point of NH_3. Justify how you make this estimate and compare with the known value of −77.7°C.
10. Look at the following trend:

Inert Gas	Atomic Number	Boiling Point (K)
He	2	4.22
Ne	8	27.07
Ar	18	87.3
Kr	36	119.93
Xe	54	165.1

This indicates that the boiling point increases with atomic number. Since inert gases interact with one another via van der Waals attraction, this trend shows that van der Waals attraction increases with atomic number. Why?

11. Look at the structure of the other two DNA bases: cytosine and guanine (Figure 1.30). Sketch how these molecules mate with each other via hydrogen bonding.

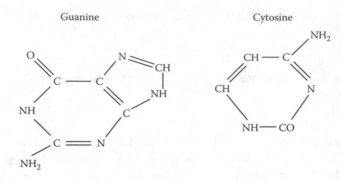

FIGURE 1.30 Molecular structure of guanine and cytosine.

12. A certain compound contains barium, titanium, and oxygen. It has a cubic unit cell structure. Barium atoms are at the corners of the unit cell, oxygen at the center of the cube faces, and titanium at the body center. Based on this information, calculate the Ba:Ti:O ratio.

13. Copper and gold at a given atomic ratio form an ordered FCC crystal structure below 380°C. Gold atoms are at the corners of the unit cell, and copper atoms are at the center of each face. Deduce the atomic ratio of copper to gold in this crystal.

14. Consider a simple cubic unit cell with side length a (i.e., having only atoms at the cube corners).
    a. Show that the packing factor is equal to $\pi/6$ for spherical atoms.
    b. There is some empty space (interstitial site) located at the center of the unit cell (0.5, 0.5, 0.5). What is the radius of the largest atom (expressed in terms of a) that can fit into this interstitial site?

15. Consider a pure amorphous carbon film with a mass density of 2.0 g/cm^3.
    a. Calculate the critical angle for total reflection at an x-ray wavelength of 0.154 nm.
    b. Beyond the critical angle, the x-ray reflected intensity exhibits oscillations as a function of angle θ due to interference of x-rays scattered from the upper and lower surfaces of the carbon film. At a film thickness of 10 nm, calculate the angular spacing $\Delta\theta$ between interference peaks.

16. The first x-ray diffraction peak obtained from nickel occurs at $\theta = 22.24°$, using an x-ray wavelength of 0.154 nm. Nickel has an FCC structure, with atomic weight of 58.7 (1 a.m.u. = 1.67×10^{-27} kg). Determine the following:
    a. Unit cell side length
    b. Atomic radius
    c. Density

17. Equation (1.8a) gives the critical angle θ_o for total x-ray reflection as a function of the x-ray wavelength λ, as repeated here:

$$\theta_o = \lambda\sqrt{\frac{nr_e}{\pi}},$$

where n is the electron density and r_e is the classical radius of the electron (2.81×10^{-15} m). Derive Equation (1.8b) to express the mass density of the material (g/cm^3) in terms of θ_o in degree, atomic number, and atomic weight using a standard copper x-ray source with $\lambda = 0.154$ nm.

18. You are given two pure copper films, A and B, which exhibit very different electrical and mechanical properties. Sample A has higher electrical resistivity and is harder than sample B. Diffraction measurements using 0.154-nm x-rays give the first diffraction peak at $2\theta = 44°$ for both samples. The width of the diffraction peak is $1.14°$ for sample A and $0.28°$ for sample B. Explain the observed difference in electrical resistivity and hardness for these two films.

2 Crystalline Imperfections and Diffusion

2.1 CLOUDY AND CLEAR ICE EXPERIMENTS

Obtain an ice cube from the refrigerator. If you inspect it carefully, you will notice that it is somewhat cloudy. That leads to two interesting questions. First, why is it cloudy? Second, how can you make clear ice cubes?

Normally, water contains dissolved air. Without dissolved air in water, fish will not survive. As water freezes to form ice, air "precipitates" out of water and escapes into the ambient. Because of the temperature gradient from the water surface to the interior, ice forms on the surface first. When this happens, air bubbles cannot escape and are trapped inside the ice cube. The cloudy appearance of the ice cube is due to the presence of these "imperfections" (air bubbles). Therefore, making clear ice cubes is then just a matter of removing as much air from the water as possible. The simplest method to remove dissolved air is to boil the water,* immediately pour the boiled water into an ice cube tray, and then place the tray inside the freezer. Without air bubbles, the resulting ice cubes will appear clear.

If you have ever been to the glaciers, you will notice the bluish color of glacier ice. Glacier ice is formed by compaction of layers of snow over thousands of years under tremendous pressure. Large air bubbles have already escaped. The resulting glacier ice is dense and absorbs long-wavelength light more efficiently. That gives rise to the blue color of glacier ice. This is the same reason why water in a swimming pool appears blue. Note that any air remaining inside the glacier ice is under large compressive stress.**

2.2 IMPERFECTIONS — GOOD OR BAD?

The preceding discussion demonstrates the effect of imperfections on properties. In Chapter 1, we introduced the argument that there is a driving force for atoms to condense into a crystalline solid. A perfect crystal represents the state of lowest potential or internal energy. Since a system of lowest potential or internal energy is *normally* considered to be the most stable, we may be led to think that all crystals should be perfect. However, at temperatures above absolute zero, each atom acquires a certain amount of thermal energy, which is responsible for the existence of imperfections.

Using the arguments of statistical thermodynamics, we can show that a system at constant volume responds to thermal fluctuations by minimizing the quantity $F = E - TS$, where E is the internal energy, T absolute temperature, and S entropy of the system (recall that entropy is a measure of the disorder of the system). F is known as the free energy of the system. Conceptually, thermal fluctuations provide some probability for the atom to be excited from the perfect lattice site, leading to the formation of crystalline defects (*structural imperfections*). As the number of defects increases, E increases (which in turn increases the free energy). This is balanced by the increase of entropy,

---

* There are at least two other methods to remove dissolved air in water: pumping on it to extract air from water or purging with helium.
** Putting a piece of glacier ice into your favorite cold drink can be hazardous to your health. As the glacier ice melts, the trapped air escapes. Since the trapped air is under tremendous pressure, the escaped air may cause your drink to explode.

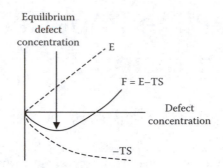

FIGURE 2.1 Free energy versus defect concentration.

resulting in the decrease of F. The system is in equilibrium when the concentration of defects results in minimum free energy for the system, as shown schematically in Figure 2.1.

Imperfections can also be introduced by the presence of impurities, as illustrated by the cloudy ice experiment (*compositional imperfections*). Silicon is probably one of the purest materials we have learned to synthesize; yet, it is difficult to make silicon with impurity content much less than 1 ppb (part per billion).

We should not consider these structural and compositional imperfections as something bad.* On the contrary, we introduce these imperfections deliberately to modify and control properties of materials. In fact, it is one of many tricks materials scientists and engineers use to obtain desirable properties of materials. For example, pure silver is too soft to be useful as utensils; the addition of a small percent of copper makes it much stronger and useful. This is known as Sterling silver. Likewise, pure aluminum and titanium are not strong enough for making aircraft components and structures**; the addition of other elements such as niobium improves strength and thermal stability. *Without the ability to introduce impurities deliberately and to control the microstructure, the microelectronics revolution and the development of advanced engineering materials would not have been possible.*

2.3 SOLID SOLUTIONS

Let us explore the issue of structural and compositional imperfections further with an example. Consider the introduction of nickel atoms into an otherwise pure copper single crystal (both elements crystallize in the FCC structure). The atomic radius of nickel (0.125 nm) is almost identical to that of copper (0.128 nm), and nickel atoms can comfortably substitute for copper atoms in lattice sites. This mixture is known as a substitutional solid solution. Cu and Ni are fully miscible. On the other hand, copper can dissolve only up to 38 a/o (atomic percent) of Zn. There are two major differences between Cu–Ni and Cu–Zn: (1) the atomic radius of Zn is about 10% larger (0.139 nm) than that of copper; and (2) Zn crystallizes in the HCP structure, while Ni crystallizes in the FCC structure.

---

* Pure and perfect diamond is colorless. When doped with impurities or irradiated by energetic electrons or gamma rays to produce defects, diamond becomes colored. Pink diamond is so rare that it is more expensive than clear diamond. Imperfections are good even in the world of precious stones.

** High-strength lightweight materials are particularly important in aviation. Consider a small aircraft such as the Cessna 172 with a maximum gross weight of 2400 lb. It is designed to have a wing loading of 13.8 lb/ft² under normal cruise conditions (the corresponding value for a Boeing 747 is about 10 times higher). As a general safety margin, wings of general aviation aircraft in the normal category are designed to withstand 3.8 *g* upward (direction of normal lift) and 1.76 *g* downward without permanent deformation after the load is removed. This means that the wings can withstand positive G-forces up to 9120 lb without damage.

Generally, mutual solubility in solid solutions is determined by four factors; in order of importance, they are *atomic size*, *crystal structure*, *electronegativity*, and *valence*.* These four factors arise from two considerations: mechanical and chemical driving forces. For example, when the atomic size of the two components differs too much (15% appears to be the magic number), there will be a large elastic strain energy penalty in dissolving substantial concentrations of solute atoms in the host lattice. When there is a large difference in electronegativity, there is a strong chemical driving force for electrons to be exchanged — that is, formation of chemical bonds and hence compounds that may not dissolve in the host lattice.

Question for discussion: How do crystal structure and valence difference affect mutual solubility? Discuss this in terms of mechanical and chemical driving forces as presented in the preceding paragraph for atomic size and electronegativity. (Hint: Remember one of the fundamental principles presented in Chapter 1: Why does a given element acquire a certain crystal structure? Will this element be "happy" if forced into a different structure? When there is a large difference in valence, this usually means that the two elements are far apart in the periodic table. How will this affect the difference in electronegativity?)

Not all impurities go into substitutional sites. For example, carbon impurities in iron go into interstitial sites — that is, sites between normal atomic positions. These interstitial sites are usually quite small. Putting impurity atoms into these sites results in substantial elastic deformation and hence a significant energy penalty. Therefore, the solubility is low for these interstitial impurities. In BCC iron, the radius of a certain interstitial site is 0.036 nm. This is significantly smaller than the atomic radius of carbon (0.075 nm). In FCC iron, the interstitial site radius is 0.053 nm, larger than that in BCC iron. Therefore, the solubility of carbon in FCC iron is larger than that in BCC iron at the same temperature.

Question for discussion: The packing factor of a BCC unit cell (0.68) is less than that of an FCC unit cell (0.74). This means that there is more empty space in a BCC unit cell. The discussion in the preceding paragraph indicates that interstitial sites in BCC unit cells are smaller than those in FCC. Resolve this apparent paradox. (Hint: The total empty space determines the packing factor, not just the size of one specific interstitial site.)

2.4 POINT DEFECTS

There are two types of intrinsic point defects: vacancies and interstitials (Figure 2.2). As discussed in the preceding section, the number of such defects is controlled by the minimization of the free energy. At thermal equilibrium, the number of vacancies N_v produced by thermal excitation is given by:

$$N_v = N \exp\left(-\frac{Q_v}{kT}\right), \tag{2.1}$$

---

* The requirement for mutual solubility in liquid solutions is that the two compounds should be chemically similar. For example, water is polar, and it readily dissolves in a polar liquid such as acetic acid (to form vinegar). On the other hand, water dissolves only slightly in nonpolar solutions such as octane. This slight solubility can be a problem for fuel systems in cars and planes. When the fuel system in a car or plane is exposed to subfreezing temperatures, water precipitates from the solution as ice crystals, which may clog fuel lines and cause engine stoppage. Therefore, special fuel additives are introduced to prevent ice crystal precipitation.

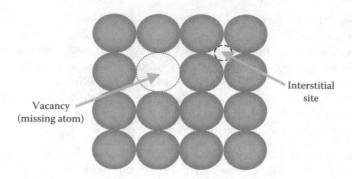

FIGURE 2.2 Vacancies and interstitials.

where N is the total number of atomic sites and Q_v the activation energy for vacancy formation (see appendix for derivation of this formula). The exponential term is known as the Boltzmann factor.*

Thermal activation also controls the migration of atoms to interstitial sites. As indicated earlier, placing atoms in interstitial sites results in significant elastic strain energy penalty. Therefore, the concentration of atoms occupying interstitial sites is typically small, unless the species involved are extremely small (e.g., atomic hydrogen). Note that in addition to thermal activation, vacancies and interstitials can be produced by other energetic means such as particle** (electrons, ions, neutrons, and x-rays) irradiation and mechanical deformation.

EXAMPLE

Consider a simple cubic unit cell, side length a, with eight identical atoms, radius R, at the corner of a cube. There is one interstitial site at the center of the cube. Calculate the radius of the interstitial site.

Solution

Consider the arrangement of atoms along the cube diagonal as shown in Figure 2.3. Let r be the radius of the interstitial site in question. The length of the diagonal is $a\sqrt{3}$. Therefore, we have:

$$a\sqrt{3} = 2R + 2r.$$

Substituting $a = 2R$, we have:

$$2\sqrt{3}R = 2R + 2r,$$

thus solving:

$$\frac{r}{R} = \sqrt{3} - 1 = 0.73.$$

---

* The Boltzmann factor appears in many phenomena involving thermal activation. The fact that the atomospheric pressure decreases with altitude can be attributed to the same factor, as shown next. A molecule of mass m at altitude h of an atmosphere has a potential energy of mgh (g acceleration due to gravity) relative to molecules at sea level. The probability for the molecule to be at altitude h, relative to molecules at sea level, is given by the Boltzmann factor, i.e., $\exp(-mgh/kT)$, where T is the atmospheric temperature. Alternatively, we can write $p(h) = p(0) \exp(-mgh/kT)$, where $p(h)$ is the pressure at altitude h. Note that this formula applies to an isothermal atmosphere only.

** Materials used in a nuclear reactor are subjected to such irradiation. As a result, vacancies are produced, resulting in volume expansion (swelling). Over time, these vacancies condense to produce macroscopic voids. Obviously, this is not a good thing to have in structures around nuclear reactors. These voids weaken the structure. Under stress, they may grow into cracks, leading to catastrophic failure.

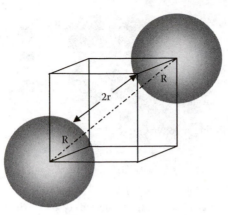

FIGURE 2.3 Interstitial in a cubic unit cell.

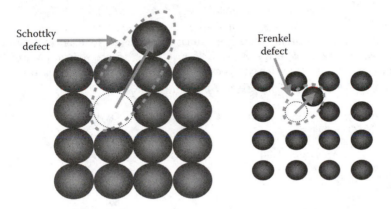

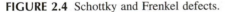

FIGURE 2.4 Schottky and Frenkel defects.

A vacancy can be formed by an atom migrating from a normal lattice site in the bulk to the surface of the crystal. This is known as a Schottky defect (Figure 2.4). Note that the formation of a Schottky defect results in volume expansion and hence density decrease. The migration of an atom from a normal lattice site to an interstitial site can also form a vacancy. This is known as a Frenkel defect (Figure 2.4). Formation of Frenkel defects does not affect density.

Point defects in ionic crystals deserve some special attention. An ionic crystal as a whole is electrically neutral. Consequently, defects in ionic crystals occur in electrically complementary pairs. For example, in a stoichiometric Na^+Cl^- crystal (i.e., the ratio of Na to Cl is exactly 1:1), when a Na^+ vacancy is produced, it can be balanced by an anion vacancy or a cation interstitial to maintain electrical neutrality. Things become a bit more interesting if stoichiometry does not occur. For example, consider titanium dioxide (TiO_2). When heated in vacuum, oxygen desorbs from the crystal, producing oxygen anion vacancies. The compound is now TiO_{2-x} ($x > 0$). If the oxygen loss is small, there is no major structural change. To maintain electrical neutrality, the two electrons associated with the oxygen anion remain, but are no longer held fixed in the oxygen site. This affects the electrical and optical properties of TiO_2. Pure stoichiometric TiO_2 is an insulator with a yellowish tint. After heating in vacuum, TiO_2 turns blue* and is electrically conducting.

---

* These oxygen vacancies arrange themselves in such a way that one can shear certain crystallographic planes of TiO_{2-x} more easily. Because of these shear planes, TiO_{2-x} functions as a solid lubricant at high temperatures.

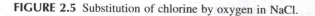

FIGURE 2.5 Substitution of chlorine by oxygen in NaCl.

EXAMPLE

Consider the introduction of oxygen into a sodium chloride crystal in which oxygen substitutes chlorine. Oxygen exists as O^{2-}. Using the principle of electrical neutrality, deduce what types of defects will be present.

Solution

Substitution of Cl^- by O^{2-} means that there is an extra negative charge. To maintain electrical neutrality, we have to remove a negative charge or add a positive charge. Since sodium positive ions are not involved in this process, the only viable response of the system is to remove a negative charge. In this case, it means the creation of an anion vacancy, as shown in Figure 2.5.

Another way to think about this is in terms of a chemical reaction. Oxygen reacts with sodium to form sodium oxide Na_2O as follows:

$$2NaCl + \frac{1}{2}O_2 \rightarrow Na_2O + Cl_2 \uparrow.$$

For each oxygen atom incorporated, one chlorine molecule is removed with one anion site occupied by oxygen and another anion site vacant.

2.5 LINE DEFECTS

Line defects can be categorized broadly into two types: edge and screw dislocations. These dislocations are formed during crystal growth or by mechanical deformation.

2.5.1 EDGE DISLOCATIONS

An edge dislocation is created when an extra plane of atoms is inserted in the middle of the crystal as shown in Figure 2.6. By convention, a positive (negative) dislocation has an extra plane of atoms in the upper (lower) plane. For a positive dislocation, atoms in the upper plane are under compression, whereas atoms in the lower plane are under tension. When given sufficient mobility, impurity atoms with an atomic size different from that of the host tend to collect at dislocations. Impurity atoms segregate to line dislocations in order to reduce the elastic strain energy. Recall that for positive edge dislocations, atoms in the upper plane are under compression. Segregation of smaller impurity atoms to the upper plane reduces the amount of compression and hence the elastic strain energy. Applying the same argument shows that larger impurity atoms favor segregation to the lower plane of a positive edge dislocation. The region around the impurity atom trapped at the dislocation core is known as the Cottrell atmosphere.

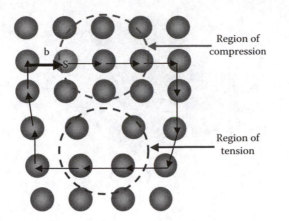

FIGURE 2.6 Structure of an edge dislocation.

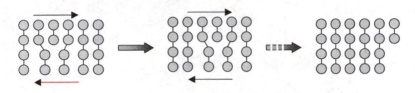

FIGURE 2.7 Applying shear stress to an edge dislocation.

The lattice distortion around the dislocation is expressed in terms of the Burgers vector b. Starting at an arbitrary lattice point (atom labeled S in Figure 2.6), we traverse a uniform clockwise circuit around the dislocation (e.g., three unit vectors to the right, three unit vectors down, three unit vectors to the left, and, finally, three unit vectors up). The lattice vector needed to close the circuit is the Burgers vector b, which is perpendicular to the dislocation line for an edge dislocation (into the plane of Figure 2.6). A positive edge dislocation moves in the direction of the Burgers vector when a positive shear stress is applied (Figure 2.7). Note that the displacement of the dislocation occurs by breaking and making one bond at a time, eventually leading to the formation of an atomic step.

2.5.2 SCREW DISLOCATIONS

One way to think about a screw dislocation is to start with a deck of cards. A "cut" is first made part way into the cards. Then the cards are sheared as shown. The resulting defect is known as a screw dislocation (Figure 2.8). The Burgers vector b as shown is a measure of the amount of distortion in generating the screw dislocation. The reason for this name is that the atoms follow a helical path around the dislocation line. Note that it is possible to have dislocations with mixed edge and screw character (i.e., an extra plane of atoms inserted along with shearing of materials in a certain direction).

2.6 PLANAR DEFECTS

Most materials are polycrystalline: They consist of single-crystal grains separated from one another through transition regions known as grain boundaries. This is the most common type of planar defect. The thickness of grain boundaries is typically a few atomic layers. Usually, there is lattice

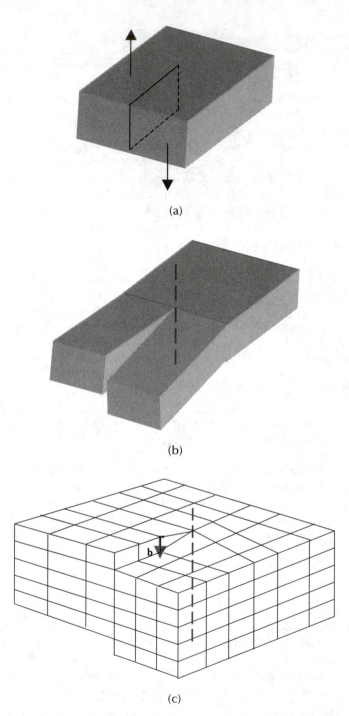

(a)

(b)

(c)

FIGURE 2.8 (a) A deck of cards to simulate the stacking of atomic planes, with a "cut" in the middle; (b) the deck is then sheared in the direction shown, resulting in the formation of a screw dislocation (The dashed line is the dislocation line.); (c) structure of a screw dislocation. The arrow shows the Burgers vector.

mismatch between adjacent grains so that the grain boundary has lower atomic packing density. This favors the segregation of large impurity atoms to grain boundaries — for example, segregation of sulfur impurities to grain boundaries in iron or nickel polycrystals.

As discussed later in the chapter on mechanical properties, grain boundaries act as obstacles to dislocation motion. Therefore, it is reasonable to deduce that grain size affects mechanical properties.* We will explore this further in that chapter. Also, grain boundaries tend to scatter electrons and phonons, thus decreasing electrical and thermal conductivity.

Another type of planar defect is the external surface of a material. Because of their unique environment, surface atoms have fewer nearest neighbors. Consequently, a surface is thermodynamically unstable and is chemically more reactive than the bulk. When given enough mobility, surface atoms will attempt to obtain maximum coordination and hence lower surface free energy. The action of catalysts (used widely in the petroleum industry and automobile catalytic converters) relies on these properties of surface atoms. Corrosion is an electrochemical process that occurs on surfaces due to heterogeneities in electrochemical potentials. Other planar defects include stacking faults (i.e., departure from the normal stacking sequence of atoms), boundaries between phases (e.g., interface between the almost pure iron matrix in steel and iron carbide precipitate), and boundaries between ferromagnetic domains.**

Measuring Average Grain Size

Given a micrograph (optical or scanning electron micrograph), the simplest way to measure the average grain size is the intercept method. Draw a line of a certain length L across the micrograph and determine the number of grain boundaries N intercepted by this line. Repeat this as many as times as practical. The average grain size is then the average of L/N.

Interestingly, the average grain boundary surface area per unit volume is equal to $2N/L$.

2.7 PRECIPITATES AS THREE-DIMENSIONAL DEFECTS

In many cases, the addition of a second component results in the formation of a compound, which then precipitates within the matrix. For example, when excess carbon is added to iron, iron carbide is formed. Materials properties depend on the number density and size of these precipitates, as well as on the interface between the precipitate and the matrix. Proper control of precipitate size and density can lead to dramatic improvements in mechanical properties due to pinning or blockage of dislocation motion.

2.8 AMORPHOUS SOLIDS

The ultimate imperfection in a solid is total disorder — an amorphous structure. Generally, an amorphous solid maintains some degree of short-range order; that is, the nearest neighbor atomic arrangements are similar to those of a crystalline solid. Beyond the nearest neighbors, an amorphous solid does not possess the long-range order of its crystalline counterpart. A common amorphous solid is fused quartz (amorphous silicon dioxide). In amorphous silicon dioxide, each silicon atom maintains a local fourfold coordination similar to crystalline SiO_2, with the Si–O bond length of 0.21 nm.

---

* Because grain boundaries act as obstacles to dislocation motion, having more grain boundaries per unit area increases the strength of materials. Experimentally, it is found that materials strength increases roughly as $d^{-0.5}$, where d is the average grain size. This is known as the Hall–Petch relationship and is one of the reasons why one is interested in the study of nanocrystalline materials, whose grain size is on the order of nanometers.

** Information stored on a hard-disk drive is in the form of magnetic domains. Think of these magnetic domains as tiny magnets pointing in one direction or the other. In a hard disk, the domains are lined up parallel to the disk surface (along the circumferential direction). The domains do not switch from one direction to the other abruptly. The thickness of the transition region is material dependent, but is on the order of 1–5 nm. This puts an upper limit on the ultimate density of magnetic storage in this format.

Another important class of amorphous solids is metallic glass. When a metallic alloy cools slowly from the melt, a polycrystalline alloy is formed. When the same alloy is cooled rapidly from the melt (quenched), at the rate of 10^5 to $10^{6\circ}$C/s or higher, the disordered structure of the liquid is preserved, resulting in an amorphous metal or metallic glass. Synthesis of the first metallic glass was first reported in 1960. Most of the initial samples were in the form of thin films or ribbons in order to achieve these large cooling rates. It was not until the early 1980s when researchers discovered ways to make metallic glass in bulk form (millimeters in size). In the early 1990s, bulk metallic glass samples based on zirconium, palladium, and rare earths with sizes of several centimeters were synthesized; they were obtainable with cooling rates as small as a few degrees Celsius per second. State-of-the-art golf clubs are made of these metallic glass alloys.

There are several striking properties of metallic glass alloys. First, they have excellent corrosion properties, primarily because of the homogeneous nature of these amorphous metals. Second, they have high strength-to-weight ratios with relatively large elastic limits.* Third, similar to glass, they are brittle below a certain temperature. Much of today's research on metallic glass alloys is to investigate ways to make them less brittle without sacrificing their other excellent mechanical and corrosion properties.

It is also possible to make amorphous semiconductors. As discussed later, semiconductors have interesting electrical and optical properties. Compared with their crystalline counterparts, amorphous semiconductors have higher electrical resistivity and more efficient light absorption.

2.9   TEMPERATURE DEPENDENCE OF DEFECT CONCENTRATION

As indicated in preceding discussions, the concentration of defects such as vacancies depends on temperature exponentially. Let us explore this in a more general way. Consider the transformation of system A to B (e.g., diamond transforming to graphite or a perfect lattice to one with vacancies). We can write:

$$\frac{d[B]}{dt} = k[A] \, , \tag{2.2}$$

where k is known as the rate constant.

In this case, the rate of formation of B is directly proportional to the concentration of A. This is known as first-order kinetics.** As shown schematically in Figure 2.9, the system A has to climb an activation energy barrier (ΔE) in order to transform into B. The quantity k is proportional to the fraction of atoms with energy ΔE above the average and is given by:

$$k = k_o \exp\left(-\frac{\Delta E}{k_B T}\right), \tag{2.3}$$

where

k_o = a material constant

k_B = the Boltzmann constant (1.38×10^{-23} J/K)

T = the absolute temperature.

---

* For several zirconium- and palladium-based metallic glass alloys, the elastic modulus is ~100 GPa, with hardness in the range of 7–9 GPa and elastic range of 1–2 %. Because of these properties, amorphous metals are being used as golf club materials for increased range.

** For systems obeying nth-order kinetics, the corresponding equation is:

$$\frac{d[B]}{dt} = k[A]^n.$$

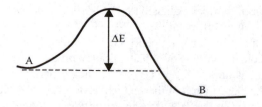

FIGURE 2.9 Activation barrier in the transformation from system A to B.

The exponential term is known as the Boltzmann factor. Because of the appearance of temperature in the exponent, the Boltzmann factor is sensitive to temperature.

EXAMPLE

It is known that the activation energy for the formation of a copper vacancy is 0.9 eV. Calculate the ratio of copper vacancy concentration at 800°C to that at 500°C.

Solution

The ratio is given by:

$$\frac{\exp(-E/k_B T_1)}{\exp(-E/k_B T_2)},$$

where

$E = 0.9$ eV
k_B = Boltzmann constant
$T_1 = 800°C$, or 1073 K
$T_2 = 500°C$, or 773 K.

Substituting, we have:

$$\frac{\exp\left(-\dfrac{0.9 \times 1.6 \times 10^{-19}}{1.38 \times 10^{-23} \times 1073}\right)}{\exp\left(-\dfrac{0.9 \times 1.6 \times 10^{-19}}{1.38 \times 10^{-23} \times 773}\right)} \approx 44.$$

That is, copper at 800°C has 44 times more vacancies than at 500°C. Assuming that vacancies are formed by moving atoms to other regular lattice sites (rather than interstitial sites), this gives rise to a larger crystal. This is one mechanism of thermal expansion.

Since producing vacancies requires the breaking of bonds, a correlation between the activation energy for vacancy formation and bond strength is expected. The preceding discussion indicates that formation of vacancies results in thermal expansion. Combining these two pieces of information, we conclude that a strong material should have a low coefficient of thermal expansion. In general, this is found to be the case for elemental solids.* For example, for a strong material such

---

* Another mechanism of thermal expansion is atomic in origin: asymmetry in the potential energy curve between atoms in solids. In crystalline solids, the coefficient of thermal expansion depends on the crystallographic direction.

as diamond, the coefficient of thermal expansion is $0.5 \times 10^{-6}/°C$ near room temperature, which is much smaller than that of Al, a soft metal ($2.5 \times 10^{-5}/°C$).*

This correlation does not work for compounds because one element of a given compound may be more volatile than the other and thus dominate vacancy production. For example, TiN is much stronger than steel, but it has almost the same coefficient of thermal expansion as steel (around $8 \times 10^{-5}/K$). Because of the close match of thermal expansion between TiN and steel, there is little or no thermal stress developed at the TiN–steel interface during temperature excursions. Thermal stress at the interface provides a driving force for coating delamination. As a result, TiN is an excellent wear-protective coating material for steel.

2.10 ATOMIC DIFFUSION

Without structural imperfections such as vacancies or interstitials, atomic diffusion in bulk solids would be minimal, and computers and advanced aerospace alloys would probably not exist. We can think of diffusion as a musical chair process, facilitated by atoms jumping into adjacent vacancies or interstitial sites. As atoms jump from one site to another, bonds have to be broken. Therefore, the activation energy for diffusion is related to bond energy. For example, aluminum has a melting temperature of 660°C and activation energy for self-diffusion (i.e., aluminum atoms diffusing in an aluminum matrix) of 165 kJ/mol. Molybdenum, a much stronger metal, has a melting temperature of 2600°C and an activation energy for self-diffusion of 460 kJ/mol. Generally, because of the small size of interstitial sites, the interstitial mechanism operates only for small atoms.

Two laws govern diffusion. Fick's first law of diffusion states that the diffusion flux J (number of atoms diffusing through per unit area per unit time) is proportional to the concentration gradient dC/dx:

$$J = -D\frac{dC}{dx}.$$ (2.4a)

Fick's second law relates the change in concentration profile as a function of time:

$$\frac{\partial C}{\partial t} = D\frac{\partial^2 C}{\partial x^2},$$ (2.4b)

where C is the concentration as a function of distance x and D the diffusivity or diffusion coefficient, given by:

$$D = D_o \exp\left(-\frac{Q}{k_B T}\right),$$ (2.5)

where

D_o = a material constant
Q = the activation energy for diffusion
k_B = the Boltzmann constant
T = absolute temperature.

---

* As benign as thermal expansion may sound, it is one of the causes of failure in microelectronics, interconnects, and packaging. For example, silicon chips are directly soldered onto ceramic carriers. The solder material has a different coefficient of thermal expansion compared with silicon and ceramics. When electronic circuits are turned on, the subsequent temperature increase results in strains at the solder joints. Over time, thermal cycling gives rise to cracks and eventual failure of these solder joints.

Q depends on the bond energy, as well as on the openness of the diffusion passageway compared with atomic size. To illustrate the latter point, note that the diffusivity of carbon atoms in BCC iron at 900°C is 1.7×10^{-10} m²/s, while diffusivity of carbon atoms in FCC iron (more closely packed) is smaller — 5.9×10^{-12} m²/s — at the same temperature. Although the interstitial sites in an FCC lattice are larger (and therefore can dissolve more interstitial atoms such as carbon), the passageways are narrower than those in BCC. Consequently, the diffusion through these passageways is slower.

In polycrystalline solids, grain boundaries become a preferred path for atomic diffusion. Because of the more open nature of grain boundaries, the activation energy for diffusion along grain boundaries is generally lower than that in the bulk.*

Derivations of the two Fick's laws are shown next.

Derivation of Fick's First Law of Diffusion

Consider diffusion across a unit area located at some arbitrary position x as shown in Figure 2.10. Assume that, in time Δt, atoms diffuse an average distance Δx along the x-direction. The flux of atoms crossing the boundary at x is given by the net difference in diffusion flux coming from the left and the right. The diffusion flux (i.e., the number of atoms per unit area per unit time) crossing position x from the left must originate between $x - \Delta x$ and x and is given by:

$$\frac{1}{4}C(x - \Delta x)\frac{\Delta x}{\Delta t} \approx \frac{1}{4}\frac{\Delta x}{\Delta t}\left[C(x) - \Delta x \frac{dC}{dx} \right]. \tag{2.6}$$

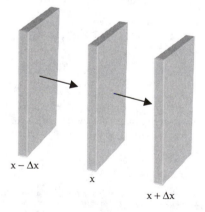

$$x - \Delta x$$

$$x$$

$$x + \Delta x$$

FIGURE 2.10 Diffusion across a boundary at x.

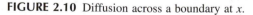

* Stress and electrical currents affect diffusion along grain boundaries. The former results in creep — continued elongation of a solid under stress with time. The latter is called electromigration — materials transport from one part of the solid to another. This was a well-known failure mode of aluminum interconnects in the early days of integrated circuits. An accident led to the discovery and widespread use of Al with 1–3% Cu as the interconnect material in the 1980s that gave much smaller electromigration rates. Pure Cu was known to have even better electromigration performance, but it was not used then because of the lack of dry-etching methods for Cu. Subsequent advances in chemical–mechanical polishing (CMP) eventually led to the use of Cu as the interconnect material of choice beginning in the mid-1990s.

The factor of 1/4 is from averaging the random arrival directions.* The second equality comes from the Taylor expansion of $C(x - \Delta x)$.** Similarly, the diffusion flux crossing position x from the right must originate between x and $x + \Delta x$ and is given by:

$$\frac{1}{4}C(x+\Delta x)\frac{\Delta x}{\Delta t} = \frac{1}{4}\frac{\Delta x}{\Delta t}\left[C(x)+\Delta x\frac{dC}{dx}\right]. \tag{2.7}$$

Therefore, the net diffusion flux $J(x)$ crossing x is equal to:

$$\frac{1}{4}C(x-\Delta x)\frac{\Delta x}{\Delta t} - \frac{1}{4}C(x+\Delta x)\frac{\Delta x}{\Delta t} \cong -\frac{1}{2}\frac{(\Delta x)^2}{\Delta t}\frac{dC(x)}{dx} = -D\frac{dC(x)}{dx}. \tag{2.8}$$

Note that diffusivity D = (diffusion distance)^2/(2 × diffusion time).*** This derivation assumes that there is conservation of materials during the diffusion process.

Derivation of Fick's Second Law of Diffusion

The derivation of Fick's second law also assumes conservation of matter. With this assumption and referring to Figure 2.10, we would like to know how the concentration at position x changes with time. The rate of change of concentration $C(x)$ depends on atom fluxes into and out of position x:

$$\frac{\partial C(x)}{\partial t} = -\frac{J(x+\Delta x)-J(x-\Delta x)}{2\Delta x}, \tag{2.9}$$

where Js represent the atom fluxes. The right-hand side of Equation (2.9) can be simplified as:

$$-\frac{J(x+\Delta x)-J(x-\Delta x)}{2\Delta x} = -\frac{\partial J}{\partial x}. \tag{2.10}$$

---

* Atoms can move in six directions ($\pm x$, $\pm y$, and $\pm z$). Therefore, the probability of moving in the $+x$ direction should be about one sixth. To calculate the exact factor, note that the effective arrival rate of atoms is proportional to cos θ, where θ defines the trajectory of the atoms relative to the surface normal, ranging from $-\pi/2$ to $\pi/2$. The average of cos θ over all possible trajectories is equal to 1/4.

** Taylor expansion of $C(x + a)$ for small a is given by:

$$C(x+a) = C(x)+\frac{a}{1!}\frac{dC}{dx}+\frac{a^2}{2!}\frac{d^2C}{dx^2}+\frac{a^3}{3!}\frac{d^3C}{dx^3}+\ldots$$

*** A historical note is in order here. In this analysis, we expand the concentration difference to first order only (i.e., up to dC/dx). Therefore, this analysis and Fick's first law implicitly assume that the concentration varies slowly with distance. What would happen when the concentration varies rapidly with distance? John Hilliard at Northwestern University raised this question during the late 1960s and early 1970s. He decided to investigate diffusion under sharp concentration gradients using superlattice coatings, i.e., coatings consisting of alternating layers of two materials. In the process, he and his colleagues discovered that when the superlattice period (repeat distance between layers) approaches 1–20 nm, these coatings attain unusual mechanical, electronic, and magnetic properties that are not simply the average of the two components. By all accounts, Prof. Hilliard unknowingly opened the field of nanoscale superlattice coatings.

Using Fick's first law of diffusion, we have:

$$\frac{\partial C}{\partial t} = -\frac{\partial J}{\partial x} = D\frac{\partial^2 C}{\partial x^2}, \tag{2.11}$$

where D, the diffusivity, is assumed to be independent of position x.

2.10.1 Diffusion due to a Step-Function Concentration Profile

As illustrated in a later section, we often modify materials properties such as hardness and electrical conductivity by diffusing impurities into the surface. Experimentally, we hold the surface concentration of the impurity constant and let impurity atoms diffuse into the material over a certain time period. The initial concentration can then be approximated by a step-function, and the concentration profile of the impurity can be obtained by solving Fick's second law, Equation (2.4b). The solution is given by:

$$\frac{C_x - C_o}{c_s - C_o} = 1 - erf\left(\frac{x}{2\sqrt{Dt}}\right), \tag{2.12}$$

where the error function, $erf(z)$, is given by:

$$erf(z) = \frac{2}{\sqrt{\pi}}\int_0^z \exp(-y^2)dy, \tag{2.13}$$

where

C_s = the surface concentration of the diffusing species
C_o = the background concentration of the diffusing species in the solid
C_x = the concentration at distance x from the initial surface at time t
D = the diffusion constant.

Table 2.1 shows the variation of $erf(z)$ versus z. Note that $erf(z) \approx 1.1\,z$ when $z \ll 1$ and $erf(z) \to 1$ as $z \gg 1$.

Figure 2.11(a) shows the error function. Figure 2.11(b) shows the impurity concentration profile at different times. Note the "spreading" of the concentration profile with time. As discussed in the next section, we take advantage of this diffusion phenomenon in forming hard surfaces on low-carbon steels and in controlling the electrical conductivity of semiconductors through controlled incorporation of impurities (i.e., doping), often via diffusion.

2.10.2 A Word about Diffusion Distance

Sometimes, it is useful to think of an average diffusion distance. In the derivation of Fick's first law, we show that the average diffusion distance is equal to $\sqrt{2Dt}$. The following example shows another derivation of the (root-mean-square) average diffusion distance. An inspection of Figure 2.11(b) shows that, at this distance, the impurity concentration decreases to about one third of the surface concentration.

TABLE 2.1
Error Function

z	erf(z)
0.1	0.1125
0.2	0.2227
0.4	0.4284
0.5	0.5205
0.6	0.6039
0.7	0.6778
0.8	0.7421
0.9	0.7970
1.0	0.8427
1.2	0.9103
1.5	0.9661
1.7	0.9838
2.0	0.9953

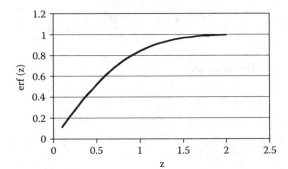

(a)

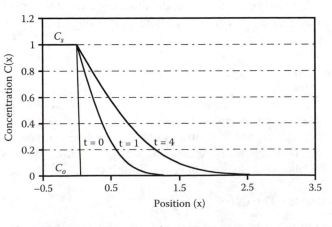

(b)

FIGURE 2.11 (a) The error function; (b) impurity concentration $C(x)$ versus position x. Arbitrary units are used in this plot, and the diffusivity is assumed to be equal to 0.1. At $t = 1$, the root-mean-square average diffusion distance $= \sqrt{2 \times 0.1 \times 1} = 0.45$. At $t = 2$, the diffusion distance $= \sqrt{2 \times 0.1 \times 4} = 0.89$.

EXAMPLE

Consider atoms moving in random directions with average speed v that are scattered after an average distance λ, known as the mean free path. It is known from statistics that after N scattering steps, the root-mean-square average diffusion distance x_{RMS} from the origin is equal to $\lambda\sqrt{N}$. Show that the root-mean-square average diffusion distance is equal to $\sqrt{2Dt}$.

Solution

The time t required to take these N steps is just $N\lambda/v$. Refer to Figure 2.10 and a system with atomic concentration C. The atoms can go left or right. Rewriting Equation (2.8), the flux of atoms at x is given by:

$$\frac{1}{2}C\left(x-\frac{\lambda}{2}\right)v - \frac{1}{2}C\left(x+\frac{\lambda}{2}\right)v = -\frac{1}{2}\lambda v\frac{dC(x)}{dx}$$

$$= -D\frac{dC(x)}{dx},$$

where λ is the average distance between collisions, or mean free path.

Therefore, we have:

$$D = \frac{1}{2}\lambda v.$$

From the statistics result, we can write:

$$v = \frac{N\lambda}{t} = \frac{x_{RMS}^2}{\lambda t}.$$

Substituting this into the preceding expression for D, we have:

$$D = \frac{x_{RMS}^2}{2t},$$

from which we obtain:

$$x_{RMS} = \sqrt{2Dt}.$$

Question for discussion: Osmosis is the diffusion of water from high to low water potential through a selectively permeable membrane (e.g., cell membranes). It is an important biological process, and it provides a method to produce potable water. It has been said that a person who is lost at sea and runs out of drinking water, should not drink sea water because it will worsen thirst or accelerate dehydration. What do you think? (Hint: Compare the salt concentration in seawater to that in the human body.)

2.11 APPLICATIONS OF IMPURITY DIFFUSION

2.11.1 CASE HARDENING

In some applications involving the use of low-carbon steels, it is important to have a hardened surface for improved wear resistance (gears and components with rotating or sliding motion) while maintaining a softer and tougher bulk for improved fracture resistance. One way to improve the surface hardness of ordinary low-carbon steels is the method of gas carburizing. In this process, the steel is exposed to a carbon-containing gas such as methane at high temperatures. The gas decomposes to produce carbon, which then diffuses and dissolves into the steel matrix, resulting in a harder surface (case). Carbon atoms can also be produced in a plasma, formed by low-pressure (10–100 mtorr) discharge of carbon-containing gases. In either case, the process is known as case hardening.

EXAMPLE

Consider a low-carbon steel with 0.1 weight percent (w/o) carbon subjected to gas carburizing. The conditions are as follows: (1) the surface carbon concentration is 1.0 w/o; (2) the diffusivity of carbon in steel at the carburizing temperature (1000°C) is 1×10^{-10} m²/s. Calculate the carbon concentration at 1.0 mm below the surface after 1 h.

Solution

The equation of interest is Equation (2.12):

$$\frac{C_x - C_o}{C_s - C_o} = 1 - erf\left(\frac{x}{2\sqrt{Dt}}\right),$$

where

C_x = the quantity we wish to obtain at x = 1.0 mm or 10^{-3} m
C_o = 0.1 w/o
C_s = 1.0 w/o.

In this example, $D = 1 \times 10^{-10}$ m²/s and t = 3600 s. Therefore, we can write:

$$\frac{C_x - 0.1}{1.0 - 0.1} = 1 - erf\left(\frac{10^{-3}}{2\sqrt{10^{-10}.3600}}\right)$$

$$= 1 - erf(0.833).$$

Using Table 2.1, we can interpolate $erf(0.833)$ to be about 0.76, which leads to C_x = 0.32 w/o.

Question for discussion: How do we control the carbon surface concentration Cs?

2.11.2 IMPURITY DOPING OF SEMICONDUCTORS

Diffusion of specific impurity atoms into a pure semiconductor is central to the development of semiconductor-based circuits and devices. We will have a more detailed discussion of this in a later chapter. The basic idea is similar to that of case hardening — that is, the semiconductor is exposed to a gas containing the impurity atom in question at a specific temperature for a specific time. One important difference is that the concentration involved is usually much lower. In case hardening of steels, the carbon concentration is usually at the percent level. In impurity doping of semiconductors, the impurity concentration typically ranges between parts per million (ppm) to parts per billion (ppb) levels.

EXAMPLE

There is another important difference in impurity doping between case hardening of steels and semi-conductors. This is best illustrated with this example. Consider the introduction of boron impurities into silicon at 1000°C. At this temperature, the diffusivity of boron in silicon is 2×10^{-18} m²/s. Assuming the surface concentration of boron C_s to be 1×10^{-6} and C_o to be zero, sketch the concentration profile of boron (i.e., boron concentration as a function of distance into the silicon substrate) after 1 h at 1000°C.

Solution

Again, the equation of interest is Equation (2.12):

$$\frac{C_x - C_o}{C_s - C_o} = 1 - erf\left(\frac{x}{2\sqrt{Dt}}\right).$$

Using this information, we can rewrite as:

$$C_x = 10^{-6}\left[1 - erf\left(\frac{x}{2\sqrt{Dt}}\right)\right].$$

Therefore,

$$C_x = 10^{-6}\left[1 - erf\left(\frac{x}{1.7 \times 10^{-7}}\right)\right] = 10^{-6}\left[1 - erf\left(\frac{x(micron)}{0.17}\right)\right].$$

We can now plot the variation of boron concentration as a function of distance into silicon, as shown in Figure 2.12. One important observation is that the boron concentration varies from about 1 to almost 0 ppm over a distance on the order of 0.2 μm. For certain semiconductor devices, 0.2 μm is considered to be too large. Another technique, known as ion implantation, is used to introduce impurities with sharp concentration profiles. In this technique, impurity species in the form of ions are accelerated to a well-defined kinetic energy and impinge onto the semiconductor. Their penetration distance is directly related to kinetic energy and has a narrower spread compared with diffusion.

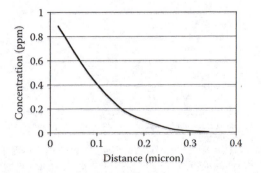

FIGURE 2.12 Hypothetical boron concentration in silicon.

2.12 DIFFUSION IN BIOLOGICAL SYSTEMS

The type of diffusion discussed previously is known as passive diffusion, which involves the movement of atoms or molecules in the direction determined by concentration gradients and without any external energy input. In biological systems such as cells, passive diffusion of species through cell membranes plays an important role — for example, gases (oxygen and carbon dioxide), hydrophobic molecules such as benzene, and small polar molecules (water and ethanol). A major component of the cell membrane is the phospholipid* double layer. These molecules dissolve in the phospholipid double layer and diffuse across the cell membrane. Osmosis — the transport of water from high to low water potential through the cell membrane or other selective permeable membranes — is an example of such passive diffusion.

Another type of diffusion occurring across cell membranes is known as facilitated diffusion. In this case, atoms or molecules to be transported do not dissolve in the phospholipid double layer. Examples include ions or charged molecules (H^+, Na^+, K^+, Ca^{2+}, Cl^-, and amino acids) and large polar molecules such as glucose. The transport occurs via two types of proteins, known as carrier proteins and channel proteins, that are present on cell membranes. Carrier proteins first bind to the species, then undergo conformational (shape) changes that allow the species to be transported through the cell membrane and released into the cell interior. This process is schematically illustrated in Figure 2.13.

Channel proteins can be considered to be open pores in the membrane that allow passage of small molecules of appropriate size and charge. Ion channels, one specific class of channel proteins, mediate the passage of ions across all membranes (Figure 2.14). They are particularly important in nerve and muscle cells, in which the opening and closing of these channels are responsible for the transmission of electrical nerve signals. The transport of ions through ion channels is fast, exceeding 10^6 ions/s** (about 1000 times faster than transport through carrier proteins); selective; and regulated by specific gates, which can be signaling molecules or electrical potential across the membrane.

2.13 WHAT IS NEXT?

Structural and compositional imperfections, in appropriate concentrations, can lead to improved mechanical properties and to important phenomena such as diffusion. They are also key to the microelectronics revolution that began in the late 1940s. Without the controlled introduction of impurities, we cannot make useful semiconductor devices, and the only time we would see silicon would be on beaches, rather than as chips in our computers or mobile phones. What are these semiconductor gizmos? How do they work? Why are impurities so important? Curious? Let us move on and explore.

---

* Phospholipids consist of glycerol, two fatty acid tails, and a charged phosphate-containing group. The charged end is hydrophilic, while the tails are hydrophobic.

** This means that the ion diffuses through the membrane (thickness ~ 10 nm) in about 1 μs. Since diffusion distance z_D is $\sqrt{(2Dt)}$, we can then write $D = z_D^2/2t$. Therefore, the diffusivity through the ion channel is equal to $10^{-16}/2 \times 10^{-6} = 5 \times 10^{-11}$ m²/s at the body's nominal temperature of 37°C. This is exceedingly fast when compared with atomic diffusion in crystalline materials at similar temperatures.

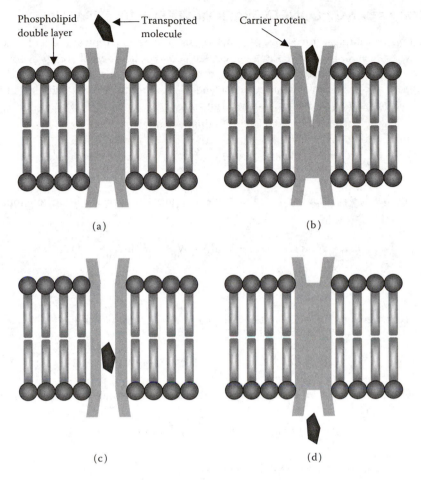

FIGURE 2.13 Facilitated diffusion via a carrier protein undergoing conformational changes from (a) to (d).

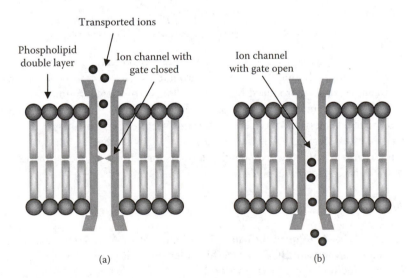

FIGURE 2.14 Facilitated diffusion via an ion channel.

APPENDIX: VACANCY CONCENTRATION VERSUS TEMPERATURE

Consider a crystal with N_{tot} lattice sites, of which N sites are occupied by atoms and N_v sites vacant. Therefore, $N_{tot} = N + N_v$. Referenced to a perfect lattice, the total internal energy E for this system is $N_v Q_v$, where Q_v is the energy required to create a lattice vacancy. The entropy S is equal to $k_B (\ln \Omega)$, where k_B is the Boltzmann constant and Ω is the number of distinguishable ways of arranging N atoms and N_v vacancies, which is given by:

$$\Omega = \frac{N_{tot}!}{N! N_v!}.$$

(2.14)

Assuming that $N \gg N_v$ and knowing that $\ln x! \approx x \ln x - x$ for $x \gg 1$, we can approximate $\ln \Omega$ as follows:

$$\ln \Omega \approx N_v \ln(N + N_v) + N \ln(N + N_v) - N_v \ln N_v - N \ln N$$

$$\approx N_v \ln N - N_v \ln N_v.$$

(2.15)

Since the free energy F for the system is equal to $E - TS$, we can write:

$$F = N_v Q_v - kT(N_v \ln N - N_v \ln N_v).$$

(2.16)

The system is in thermal equilibrium when the free energy is minimum (i.e., $dF/dN_v = 0$):

$$0 = Q_v - kT(\ln N - \ln N_v - 1) \approx Q_v - kT(\ln N - \ln N_v).$$

(2.17)

Simple manipulation of Equation (2.17) gives:

$$N_v = N \exp\left(-\frac{Q_v}{kT}\right).$$

(2.18)

In deriving this formula, we assume that vacancies are produced by the migration of atoms to other regular atomic sites with identical atomic environments. However, this is not always the case. For example, vacancies can be produced by moving atoms from regular atomic sites to interstitial sites. In that case, the configuration entropy calculated here has to be modified. Furthermore, atoms in interstitial sites most likely have different vibration frequencies and hence different entropies associated with atomic vibrations. The net result is that when vacancies are formed by migration of atoms to interstitials, only the exponential term is preserved in Equation (2.18). The pre-exponential term is no longer equal to N.

PROBLEMS

1. Fe, Cr, and Ni have atomic radii within 1% of each other. Which element, Cr or Ni, will have greater solubility in BCC iron? Which element will have greater solubility in FCC iron? Please include a short explanation.

2. A single crystal of Cs^+Cl^- (BCC) is heated in vacuum, causing the escape of equal numbers of cesium and chlorine atoms. Explain which type of defect (Schottky or

Frenkel) will dominate. (Hint: Compare the size of interstitial sites with the radius of Cs^+ [0.169 nm] and Cl^- [0.181 nm].)

3. Given a crystal of sodium chloride (NaCl), we introduce small amounts of magnesium chloride ($MgCl_2$). You are given the following facts: (1) Magnesium occupies sodium sites substitutionally; (2) note the charged states: Na^+, Cl^-, Mg^{2+}; (3) electrical neutrality is always maintained for the entire crystal. Explain which type of defect (cation or anion vacancies) will be introduced.

4. The following is a table of activation energy Q for self-diffusion and sublimation energy H (i.e., energy required to transform atoms from the solid state into the vapor state) for several elements:

Element	Q (kJ/mol)	H (kJ/mol)
Al	165	284
Cu	196	304
Fe	240	354
Mo	460	535
Ni	293	380
Zn	92	115

   a. Make a plot of Q versus H. Draw the best straight line through the points and the origin. Estimate the slope of the straight line.
   b. You may notice that Q and H are highly correlated. Discuss the origin of this correlation.
   c. Note that H is always greater than Q. Speculate why this is so.

5. As the temperature is raised from 800 to 1000 K, the unit cell size of a given metal crystal (as measured by x-ray diffraction) increases 0.05% due to thermal expansion (i.e., increased equilibrium distance between atoms). In the same temperature range, the density decreases by 0.16%, due to the combination of thermal expansion and vacancy production.

   a. What is the expected percentage decrease in density due to normal thermal expansion alone? (Hint: When the unit cell size increases by $x\%$ ($<<100\%$), the volume per unit cell increases by $3x\%$.)
   b. Evaluate the net density decrease due to vacancy production alone.
   c. If the vacancy concentration at 800 K is 10^{-6}, estimate the activation energy for vacancy production.

6. A TiO_2 crystal is grown with some Cr_2O_3 impurities. The valence of Ti is +4, and the valence of Cr is +3. Chromium ions occupy titanium sites substitutionally. Explain what type of vacancy will be introduced.

7. Consider an FCC unit cell with lattice constant a and atoms with radius R.

   a. You may notice that there is empty space along the edge of the cube, as shown in Figure 2.15. If the radius of the largest atom that can be accommodated in the cube

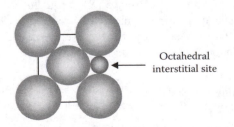

Octahedral interstitial site

FIGURE 2.15 An octahedral interstitial site in an FCC unit cell.

edge position without distorting the unit cell is r, show that r/a is equal to 0.146. Express r in terms of R. This is known as an octahedral site because it is coordinated to its six nearest neighbors in an octahedral geometry.

b. Show that there are four octahedral sites per unit cell.*

8. Consider a BCC unit cell with lattice constant a and atoms with radius R.

a. You may notice that there is empty space centered at (1/2,1/4,0). If the radius of the largest atom that can be accommodated at this location without distorting the unit cell is r, express r in terms of a and R. This is known as a fourfold site because it is coordinated to four neighbors.

b. Show that there are six such sites per unit cell.**

9. You are given three aluminum samples of the same purity: single crystal, polycrystal with an average grain size of 10 μm, and polycrystal with an average grain size of 1 μm. Now you are to measure the self-diffusion coefficient (i.e., diffusion of aluminum atoms in aluminum) for each of these three samples at the same temperature.

a. Rank the three diffusion coefficients. Explain how you arrive at this conclusion.

b. If you repeat these measurements with another metal having higher sublimation energy (e.g., copper), what will you find about the corresponding diffusion coefficients (larger or smaller)? Why?

10. In the growth of single-crystal sapphire (Al_2O_3), TiO_2 impurities are introduced. The valence of Al is +3, and the valence of Ti is +4. Titanium ions occupy aluminum sites substitutionally. Electrical neutrality is always maintained for the entire crystal.

a. Explain which vacancies (cation or anion vacancies) will be introduced.

b. For every 100 Ti ions introduced, determine the number of (cation or anion) vacancies produced.

11. Ferrous oxide (Fe_xO) has the same crystal structure as NaCl. The nominal valence state of iron in this structure is +2, with a small fraction in the +3 state. The valence state of oxygen is –2. Show that cation vacancies must exist under these conditions. Argue why $x < 1$.

12. Diffusion as a random walk problem:

a. Consider a particle executing N diffusion steps in random directions, each of length λ, which is sometimes known as the mean free path. The position of the particle R is given by:

$$\vec{R} = \vec{s}_1 + \vec{s}_2 + \ldots + \vec{s}_N,$$

where all s_is represent the individual steps. Taking the square on both sides, we have:

$$R^2 = \sum_{i=1}^{N} s_i^2 + \frac{1}{2} \sum_{i \neq j} \vec{s}_i \cdot \vec{s}_j .$$

Show that the average diffusion distance $|R|$ is equal to $\lambda \sqrt{N}$.

b. A bottle of perfume is open in the middle of a room. Someone 10 m away can smell the perfume in 1000 s. Assuming that the average speed of perfume molecules is 100 m/s, estimate the value of λ of these perfume molecules. You will find that the

---

* Another type of interstitial site is present in an FCC unit cell. There are eight per unit cell, located at (1/4,1/4,1/4) and equivalent positions. These are known as tetrahedral sites because each is coordinated to its four nearest neighbors in a tetrahedral geometry.

** Another type of interstitial site is present in a BCC unit cell. There are six per unit cell, located at (1/2,1/2,0), (0,0,1/2), and equivalent locations. These are known as octahedral sites.

computed mean free path is much greater than typical values for most molecules (10–100 nm). This reason is due to air currents in a typical room.

13. Consider the gas-carburizing problem for a steel sample with 0.2 w/o carbon. The gas carburizing is performed at 1000°C with the diffusivity of carbon in steel equal to 1×10^{-10} m^2/s, and the surface carbon concentration is 0.8 w/o. Calculate the carburizing time required to obtain 0.6 w/o carbon at 1.0 mm below the surface.

14. It is often said that atoms with lower activation energy for diffusion tend to diffuse faster than those with higher activation energy (e.g., the diffusion of atomic hydrogen and carbon in an iron lattice). Let us explore this in more detail. Consider two diffusing species in a given system, one with higher activation energy, Q_1, than the other, Q_2. Figure 2.16 shows the Arrhenius plot — that is, log(diffusivity) versus $1/T$ (T absolute temperature) — for these two species:
    a. Which is the curve with activation energy for diffusion Q_1? Which one with Q_2?
    b. From Figure 2.16, discuss the comment made at the beginning of this question.

15. When two metal components are pressed against each other inside a vacuum system and subjected to elevated temperatures (a few hundred degrees centigrade), seizure sometimes occurs (i.e., the two metal components fuse together). Explain what may have occurred. Discuss at least two strategies to minimize the occurrence of such seizures.

16. Discuss the reasons why (a) silver and gold are fully miscible; and (b) silver and molybdenum have little or no mutual solubility in each other.

17. Look at the error function. Using the impurity diffusion example given in the text, discuss how the sharpness of the diffusion profile depends on the temperature. Derive the functional relationship between temperature and sharpness of the diffusion profile.

18. Osmosis involves the diffusion of water molecules from higher to lower concentration (C_H to C_L) through a selectively permeable membrane of thickness t. According to Fick's law, the diffusion flux should be equal to the diffusion constant times the concentration gradient. Since the concentration gradient is equal to $(C_H - C_L)/t$, the diffusion flux should be inversely proportional to the membrane thickness t. The experimental result is shown in Figure 2.17. Explain the deviation of experimental data from the expected trend for small membrane thicknesses. (Hint: Think about thickness uniformity of the membrane.)

19. In deriving Fick's first law, we write the concentration as a function of position using Taylor expansion retaining only first-order terms.
    a. Define the condition under which this approach is valid.
    b. Under the condition when third-order terms have to be used, rederive the Fick's first law.

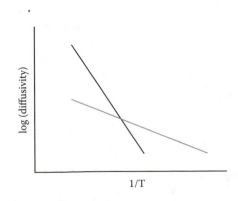

FIGURE 2.16 Arrhenius plot for two processes with different activation energies.

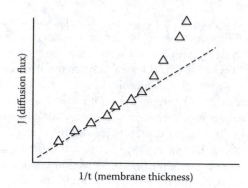

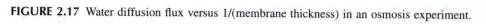

FIGURE 2.17 Water diffusion flux versus 1/(membrane thickness) in an osmosis experiment.

3 Electrical Properties of Metals and Semiconductors

3.1 WORLD OF ELECTRONICS

The invention of the transistor in 1947 marked the beginning of an unprecedented revolution that affected human activities at every level. In those days, the entertainment center for an average home was a vacuum-tube radio. It was necessary to wait several minutes for warm-up (and miss what was being broadcast) after it was switched on. The transistor radio changed that because it could be turned on instantly. For folks experiencing this change during the 1950s, this was magic. More important, because of its smaller size and low power consumption, a transistor radio could be powered by batteries and was easily portable — beach parties could not possibly have happened without these transistor radios.

Shortly afterwards, it was realized that components such as transistors and resistors did not have to be fabricated individually and soldered together on a circuit board one at a time. Instead, the pioneers of integrated circuit technology learned that it was possible to make an entire circuit on a small silicon chip, using sequential optical masking (photolithography), dopant diffusion, etching, and interconnect deposition. In fact, many hundreds of the same circuit could be made on one silicon wafer, thus reducing cost substantially. That launched the development of integrated circuits and the beginning of the microelectronics revolution.

Year after year, engineers managed to pack more transistors and circuit elements onto the silicon chip. We now have microprocessors and microcontrollers with remarkable computing powers. For example, in the late 1970s when the first personal computer became commercially available, the central processing unit (CPU), such as the Mostek 6502 (that powered the Apple II) and Intel 8080, could process about half a million instructions per second. In 2003, we have microprocessors with over ten million transistors packed into a few tens of square millimeters of silicon that can process more than one billion instructions per second.*

These fast-paced changes over the past 50 years have affected our daily lives in a profound way. Today, scientists and engineers perform computations with electronic calculators and portable computers, rather than slide rules and room-sized computers. We play music not through vacuum-tube-powered electronics, but by compact solid-state lasers. Modern flight simulators are so realistic that pilots can log instrument flying hours on them. Television images no longer appear only on supersized cathode ray tubes, but also on picture-frame-like panels lit up by solid-state devices of different variants. A car is more than an internal combustion engine with four wheels. It has become an extension of the living room and office with a dazzling array of microprocessors and microcontrollers that perform various functions: engine ignition and monitoring, emission and environmental control, entertainment, navigation, antilock brakes, deceleration sensor and airbag deployment, traction control, etc. We can now purchase handheld global positioning satellite system (GPS) receivers that capture and decode signals from a constellation of 24 satellites to obtain positions accurate to better than 10 m (Figure 3.1). These GPS receivers are ideal navigation tools for cars,

---

* An early generation CPU such as the Mostek 6502 ran at a clock rate of 1 MHz. A typical instruction (e.g., putting 1 byte in a certain register or memory location) takes between two and four clock cycles. In 2003, a late model Pentium processor runs at 2 GHz. With look-ahead and parallel processing techniques, it can execute one instruction, on average, in less than two clock cycles.

FIGURE 3.1 A handheld GPS receiver.

FIGURE 3.2 Aboard the flight deck of a Boeing 747-400.

ships, and airplanes.* With the aid of advanced avionics, a modern aircraft can fly from point A to point B, and then land automatically under zero visibility conditions (Figure 3.2).**

We are no longer bound by spatial separation in our quest to communicate. Voice and data can be transmitted and received almost anywhere in the world. We can easily talk to friends and family

---

* Handheld GPS navigation units with 12-channel parallel receivers and street-level base maps could be purchased for about $200 in the United States in 2005.
** With most standard instrument landing systems (known as category I ILS), one can land an aircraft under low-visibility conditions (200 ft ceiling and 1/2 statute mile flight visibility). Category 3c ILS, first developed for London's Heathrow Airport because of its frequent foggy weather, allows an aircraft to land with 0 ft ceiling and zero visibility. Special equipment and certification are required to perform such a landing.

via mobile phones as if they live next door. We keep imagery of significant events in megapixel digital photographs* or gigabyte movie files and store them in compact or digital video disks, rather than as paper photographs in albums or celluloid reels.

Even after more than 50 years of exponential growth, the pace of progress with electronics does not appear to have slowed down. We continue to see the invention of new electronic devices that are smaller, faster, and cheaper and that have more functionality. The development of molecular materials has spawned a new industry for electronics and display. As the feature size in these electronic circuits approaches nanometer dimensions, quantum effects become important. We expect to discover new phenomena from which novel devices and materials are certain to be developed. It will be a brave new world indeed.

3.2 DEFINITIONS AND UNITS

It is assumed that readers are familiar with definitions of electrical conductivity, resistivity, current density, and electric field, as well as Ohm's law and common units such as volt, ohm, and amp (please refer to the appendix of this chapter). Silver is an example of a good electrical conductor with electrical conductivity at room temperature equal to 6.3×10^7 $(\Omega\text{-m})^{-1}$, while diamond is an example of a poor electrical conductor with electrical conductivity less than 10^{-14} $(\Omega\text{-m})^{-1}$ at room temperature.

3.3 CLASSICAL MODEL OF ELECTRONIC CONDUCTION IN METALS

Consider a metal with n conduction electrons per unit volume, to which an electric field E is applied. This causes an acceleration of electrons equal to eE/m, where e is the electron charge and m its mass. With τ as the average time between collisions (mean free time or relaxation time),** the average drift velocity v_d acquired by electrons is equal to:

$$v_d = \frac{eE}{m}\tau. \tag{3.1}$$

This increase in the average speed of electrons*** implies that they are excited to a higher energy level. Though this sounds trivial, it is important to note that electrical conduction cannot occur without such excitations. The resulting current density J is given by:

$$J = nev_d = \frac{ne^2\tau}{m}E, \tag{3.2a}$$

where n is the conduction electron concentration. Equation (3.2a) is one form of Ohm's law, which states that the current density J is proportional to the electric field E. The constant of proportionality is the electrical conductivity σ, given by:

---

* A 5-Mpixel digital image, when presented on a standard 8 in. × 10 in. print, has an equivalent pixel size measuring approximately 100µ × 100µ, which is at the resolution limit of the unaided human eye.

** Electrons lose energy by colliding with phonons (lattice vibrations), impurities, and each other. Typical relaxation times for metals are on the order of 10^{-14} s.

*** For metals in typical situations, the drift velocity is in the range of 1–10 cm/s. This is much smaller than the thermal speeds of electrons at room temperature (~10^7 to 10^8 cm/s).

$$\sigma = \frac{ne^2\tau}{m}$$

$$= ne\left(\frac{e\tau}{m}\right) \tag{3.2b}$$

$$= ne\mu,$$

where μ $(= e\tau/m)$ is known as the mobility. Electrical resistivity ρ is defined as the reciprocal of the conductivity.

Generally, the electrical resistivity of metals increases with temperature. For most metals, the electrical resistivity increases roughly 0.4%/K. This phenomenon is highly reproducible and is the basis of using platinum thermometers in precision temperature measurements. Another way of expressing this phenomenon is that the relaxation time or electron mobility decreases with increasing temperature. The addition of impurities (e.g., in the formation of solid solutions) increases the electrical resistivity.*

EXAMPLE

It is known that gold has a density of 19.3 g/cm³. The atomic weight is 197. Each gold atom provides one conduction electron. The electrical resistivity is 2.2×10^{-8} ohm-m. Calculate the relaxation time and electron mobility in gold.

Solution

The electrical resistivity ρ is given by the following formula:

$$\rho = \frac{1}{\sigma} = \frac{m}{ne^2\tau},$$

where

σ = the electrical conductivity
n = the conduction electron concentration
e = the electron charge $(1.6 \times 10^{-19}$ C)
m = electron mass $(9.1 \times 10^{-31}$ kg)
τ = the relaxation time

In this problem, in order to calculate the relaxation time, we need to determine the conduction electron concentration. Given the density and atomic weight, we can compute the number of gold atoms per unit volume as:

$$\frac{19.3}{197} \times 6 \times 10^{23} = 5.88 \times 10^{22}/\text{cm}^3 = 5.88 \times 10^{28}/\text{m}^3.$$

---

* This statement does not apply to cases where ordered alloys such as Cu_3Au are formed, in which copper and gold atoms are sitting in well-defined lattice sites.

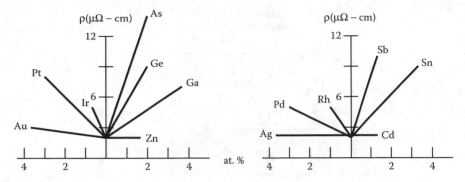

FIGURE 3.3 Resistivity versus concentration for dilute alloys of copper. (Adapted from Linde, J. O. 1932. *Annals of Physics* 15:219.)

Since each gold atom contributes one conduction electron, this is also the conduction electron concentration. Substituting this into the preceding expression for electrical resistivity, we have:

$$\tau = \frac{m}{ne^2\rho} = \frac{9.1\times10^{-31}}{5.88\times10^{28}\times(1.6\times10^{-19})^2\times2.2\times10^{-8}}$$

$$= 2.75\times10^{-14}\,\text{s}.$$

The electron mobility μ is given by:

$$\mu = \frac{e\tau}{m} = \frac{1.6\times10^{-19}\times2.75\times10^{-14}}{9.1\times10^{-31}}$$

$$= 4.84\times10^{-3}\,\text{m}^2/\text{V}-\text{s}.$$

This example shows that the average time between collisions in metals is on the order of 1×10^{-14} s. Given that the average speed of conduction electrons in metals is ~1×10^8 cm/s, the mean free path is ~$10^{-14}\times10^8 = 10^{-6}$ cm, which is about 100 times the atomic spacing.

3.4 RESISTIVITY RULES FOR DILUTE METALLIC ALLOYS

3.4.1 NORDHEIM'S RULE

In dilute binary alloys $A_{1-x}B_x$, the resistivity can be shown to vary as $(1-x)\,x$. For small x, the resistivity is proportional to the concentration of B (Figure 3.3). If this rule is valid for the entire composition range, this implies that the maximum resistivity occurs at $x = 0.5$. This rule does not apply when phases of different structures or compositions are formed as a function of x.

3.4.2 LINDE–NORBURY RULE

When different impurities are introduced into an alloy, the resistivity increase is found to be proportional to the valence difference between the impurity and the host, as shown in Figure 3.3. Similarly, this rule applies only when these impurities do not cause microstructural changes or the precipitation of new phases.

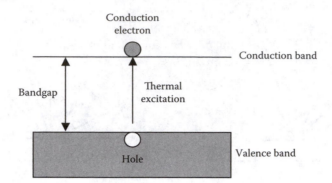

FIGURE 3.4 Bandgap in a semiconductor and thermal excitation of electrons across the bandgap.

3.5 ENERGY BAND MODEL FOR ELECTRONIC CONDUCTION

As presented in our discussion of metallic bonding, when we bring N sodium atoms together to form a solid, the N 3s levels form N closely spaced levels, known as an energy band. Since there are N 3s electrons and each level can accommodate two electrons, this means that the energy band is half filled. Individual levels within the band are so closely spaced that it takes an infinitesimal amount of energy to excite electrons from occupied to empty states.* As noted earlier, if electrical conduction is to take place, it is necessary to excite a conduction electron to an empty electronic state. The infinitesimally small spacing between occupied and empty electronic levels in metals explains why metals are good electrical conductors.

This energy model explains why all group I (A and B) elements are electrical conductors. It is left as an exercise to the readers to argue that all group III elements are also electrical conductors.** For group IV elements, the bands are full, and there is an energy gap between the highest occupied energy band (valence band) and the lowest unoccupied band (conduction band), as shown in Figure 3.4. This bandgap is about 6 eV for carbon (diamond), 1.1 eV for silicon, and 0.67 eV for germanium at room temperature. The size of the bandgap can be related to bond strength. In order to produce a conduction band electron (i.e., exciting the electron across the bandgap), an electron must be broken loose from the covalent bond. The vacancy left behind in the valence band is known as a hole (Figure 3.4).

3.6 INTRINSIC SEMICONDUCTORS

Intrinsic semiconductors (i.e., with no impurities) such as silicon have electrical conductivity between metals and insulators. Table 3.1 lists the bandgap values for several typical semiconductors. Table 3.2 compares selected properties of silicon and gallium arsenide, the two most important semiconductors in use today.

Since electrons and holes are present in a semiconductor, both contribute to electrical conduction. The current density J is then given by:

$$J = (ne\mu_n + pe\mu_p)E, \tag{3.3}$$

---

* To appreciate how closely spaced these energy levels are, consider 1 g of sodium. This piece of sodium contains about 2.6×10^{22} atoms and the same number of electronic levels within the valence band. The width of the valence band in sodium is about 10 eV. Therefore, the average energy spacing between these levels is roughly $10/(2.6 \times 10^{22}$ eV), which is roughly 3.8×10^{-22} eV. This is much less than the thermal kinetic energy of electrons at room temperature ($3/2\, k_B T \approx 0.04$ eV).
** It is beyond the scope of this book to go into the details of energy band theory for two- and three-dimensional solids. In one dimension, group II elements should be insulators (or poor conductors). However, because of band overlaps along different crystallographic directions in two- and three-dimensional solids, group II elements are electrical conductors.

TABLE 3.1
Bandgap Values for Selected Semiconductors

Material	Bandgap at 300 K (eV)
C (diamond)	6.0
GaN	3.37
SiC (4H)	3.25
SiC (6H)	3.0
AlAs	2.15
GaAs	1.4
Si	1.1
Ge	0.67
InSb	0.17
Sn (gray Sn)	0.0

TABLE 3.2
Comparison of Selected Properties of Silicon and Gallium Arsenide at 300 K

	Si	GaAs
Atomic weight	28.1	144.6
Lattice constant (nm)	0.543	0.565
Density (g/cm³)	2.33	5.32
Energy gap (eV)	1.11	1.40
Intrinsic carrier concentration (/cm³)	1.45×10^{10}	9×10^{6}
Intrinsic mobilities (cm²/V-s): electrons	1350	8600
Intrinsic mobilities (cm²/V-s): holes	480	250
Dielectric constant	11.7	12
Melting point (C)	1415	1238
Thermal conductivity (watt/cm-K)	1.5	0.81
Specific heat (J/g-K)	0.7	0.35
Thermal expn. coeff.(/K)	2.5×10^{-6}	5.9×10^{-6}

where
n and p = the electron and hole concentration, respectively
μ_n and μ_p = the electron and hole mobility, respectively

In general, $\mu_p < \mu_n$ (refer to Table 3.2).

EXAMPLE

Calculate the room temperature electrical conductivity of pure or intrinsic silicon. The electron mobility is 0.135 m²/V, and the hole mobility is 0.048 m²/V.

Solution

From Equation (3.3), the electrical conductivity σ of a semiconductor is given by:

$$\sigma = ne\mu_n + pe\mu_p.$$

Substituting the appropriate numbers in the preceding expression, we have:

$$1.45 \times 10^{16} \times 1.6 \times 10^{-19} \times 0.135 + 1.45 \times 10^{16} \times 1.6 \times 10^{-19} \times 0.048 = 4.25 \times 10^{-4} (\text{ohm} - \text{m})^{-1}.$$

Because conduction electrons are produced by thermal excitation, the conduction electron concentration must depend on temperature. This dependence can be derived as follows. First, we write a symbolic equation to represent the excitation of electrons from the valence band to produce electrons in the conduction band and holes in the valence band:

$$VB \leftrightarrow \text{electrons} + \text{holes}, \tag{3.4}$$

where VB is the semiconductor valence band.

Treating this as a reaction at thermal equilibrium, we can write:

$$\frac{[\text{electrons}][\text{holes}]}{[VB]} = C \exp\left(-\frac{E_g}{k_B T} \right), \tag{3.5}$$

where

[electrons] = electron concentration n
[holes] = hole concentration p
[VB] = total valence electron concentration
C = some constant

The other terms have their usual meanings. At constant temperature, we have:

$$np = C' \exp\left(-\frac{E_g}{k_B T} \right), \tag{3.6}$$

where C' is some constant. This means that at constant temperature, the product np is a constant for a given semiconductor. For intrinsic semiconductors, $n = p = n_i$, where n_i is the intrinsic carrier concentration. Therefore,

$$np = n_i^2 = C' \exp\left(-\frac{E_g}{k_B T} \right); \tag{3.7a}$$

$$n_i \propto \exp\left(-\frac{E_g}{2 k_B T} \right). \tag{3.7b}$$

The electrical conductivity for metals decreases with increasing temperature, primarily due to the decrease of electron mobility μ or relaxation time τ. For semiconductors, two factors control electrical conductivity: mobility and carrier concentration. For intrinsic semiconductors, the mobility is proportional to $T^{-3/2}$, while the carrier concentration n depends on temperature exponentially, as shown in Equation (3.7b). The second term is dominant at and above room temperature for typical semiconductors.

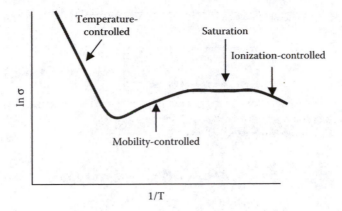

FIGURE 3.5 Variation of electrical conductivity versus temperature for a semiconductor.

As shown in Figure 3.5, at low temperatures, the electrical conductivity is controlled by charge carriers from residual impurities. At intermediate temperatures (just beyond the saturation region), the mobility term causes the conductivity to decrease. With a further increase in temperature, the exponential term begins to dominate. From this plot, we can directly determine the bandgap of the semiconductor using Equation (3.7b).

We can realize several immediate applications for semiconductors based on the existence of the bandgap:

- Thermistors. Because of the exponential dependence of conductivity on temperature, temperature sensors based on semiconductors are much more sensitive than those based on metals. Applications include temperature measurements and remote sensing.
- Photodetectors. When light of energy $h\nu \geq E_g$ shines on a semiconductor (bandgap = E_g), the concentration of charge carriers (electrons and holes) increases due to photoexcitation across the bandgap, resulting in higher electrical conductivity. Applications include communications, remote sensing, imaging, and photovoltaics.
- Light emitters. For certain semiconductors, it is possible to obtain the reverse of photoexcitation, i.e., electrons and holes recombine to produce light. In other words, electrical energy can be converted into light (in many cases, quite efficiently). Extension of this concept allows production of lasers based on semiconductors. See further discussion in Section 3.8.

Note that semiconductors can also be compounds. Some of the better known compound semiconductors include silicon carbide (better high-temperature stability than silicon), gallium arsenide (higher carrier mobilities, better light absorption and emission characteristics than silicon), gallium nitride (emitting light in the blue region), and zinc oxide (possible UV solid-state laser material).

3.7 EXTRINSIC SEMICONDUCTORS

When impurity atoms are introduced into a given semiconductor, they change the electrical properties. One key phenomenon is that impurities determine whether electrons or holes carry the bulk of the electrical current, as discussed next.

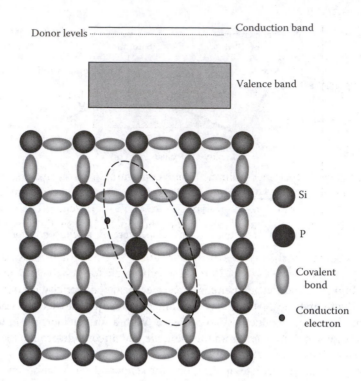

FIGURE 3.6 Donor levels in a semiconductor.

3.7.1 N-Type Semiconductors

Starting with pure silicon, consider the introduction of group V impurities such as phosphorus.* The size of phosphorus atoms is such that they occupy normal silicon lattice sites. In this case, phosphorus is known as a substitutional impurity. Since phosphorus is a group V element, it has *one more valence electron than silicon.* When P occupies substitutional sites in Si, only four valence electrons are needed to satisfy bonding in a normal Si lattice site. The extra electron is therefore weakly bound and can easily be excited to the conduction band.** This is shown schematically in Figure 3.6. The energy level occupied by this electron is known as the donor level, and phosphorus is called a donor impurity. Note that the donor level in this case is not too far from the conduction band. This illustrates the easy excitation of electrons from the donor level to the conduction band.

For most group V impurities in silicon, the observed energy separation between the donor level and the conduction band is 40–50 meV. Compared with the available thermal energy at room temperature ($3/2\ k_B T$ at room temperature $\cong$ 40 meV), this means that almost all donor atoms are ionized around room temperature, each providing one conduction electron. Since there are 5×10^{22}

---

* Impurities such as phosphorus are introduced by diffusion or ion implantation. Diffusion is relatively easy to do, but it is difficult to make an abrupt profile using diffusion. In ion implantation, impurity species in the form of ions are accelerated to a well-defined kinetic energy and impinge onto the semiconductor. Their penetration distance is directly related to kinetic energy and has a narrower spread compared with diffusion. Defects such as vacancies and interstitials are produced during the ion implanation process (ion energies on the order of tens of keV or higher are used) and are removed by annealing.
** Think of the binding of the extra electron to the phosphorus impurity site as analogous to a hydrogen atom. The binding energy of the 1s electron in hydrogen is 13.6 eV. To calculate the electron binding energy in this case, we have to make two corrections. First, the effective mass of the electron is lower in silicon by a factor of ~3–5, due to the interaction between conduction electrons and the crystal potential. Second, since the conduction medium is silicon, the binding energy has to be corrected by a factor of 1/square of the dielectric constant of silicon (11.7). This gives a theoretical estimate of the binding energy of 20–30 meV, compared with experimental values in the 40–50-meV range.

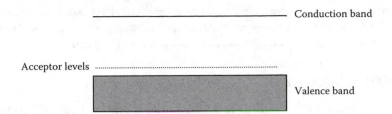

FIGURE 3.7 Acceptor levels in a semiconductor.

silicon atoms/cm³, 1 ppm (part per million) donor concentration will give rise to an extra 5×10^{16}/cm³ conduction electrons to the material. This is much greater than the silicon intrinsic conduction electron concentration at room temperature (1.45×10^{10}/cm³). Therefore, at moderate donor impurity levels, the conduction electron concentration n is essentially equal to the donor concentration N_D.

This also shows that small amounts of donor atoms can have large effects on the carrier concentration and hence electrical conductivity of a semiconductor. As noted from Equation (3.6), $np = n_i^2$ is a constant at a given temperature. Therefore, the hole concentration p is equal to n_i^2/N_D (since n is equal to N_D). With $N_D \gg n_i$, this means that the conduction electron concentration n is much greater than the hole concentration p; *electrons are the majority charge carriers and holes the minority charge carriers.* As a result, Si doped with group V impurities is known as an *n*-type semiconductor.

3.7.2 *P*-Type Semiconductors

Consider the analogous case of introducing group III impurities into silicon, such as boron. Boron occupies the normal lattice sites of silicon. Boron has three valence electrons, *one short of the four valence electrons* needed to satisfy bonding in normal silicon lattice sites. Therefore, an electron vacancy exists around each boron atom. This vacancy is known as the hole. It takes relatively little energy to excite valence electrons into this level, known as the acceptor level, as shown in Figure 3.7. Boron is called an acceptor impurity. Again, note that the acceptor level is not too far from the valence band to indicate the easy excitation of electrons from the valence band to the acceptor level.

Analogous to the case of *n*-type semiconductors, almost all acceptor atoms are occupied (i.e., charged negatively) around room temperature, each providing one hole in the valence band. In this case, the hole concentration is much greater than the conduction electron concentration, so *holes are the majority charge carriers and electrons minority carriers.* Si doped with group III impurities is known as a *p*-type semiconductor.

It is important to remember that the dopants become ionized after they accept or donate electrons to the valence or conduction band. These immobile ionic charges compensate for the mobile charges (electrons or holes), thus maintaining charge neutrality in the semiconductor.

Questions for discussion. Consider the doping of GaAs, a III–V semiconductor, by Si impurity atoms. There are two possibilities:

1. (a) If Si goes into Ga sites, what is the carrier type? (b) If Si goes into As sites, what is the carrier type? (Hint: Determine whether there is an extra electron with the substitution. If so, it will be a donor impurity.)
2. Titanium dioxide (TiO_2) is a large-gap semiconductor.* When heated in vacuum, small amounts of oxygen desorb so that the composition becomes TiO_{2-x} ($x > 0$) — that is,

---

* The bandgap of titanium dioxide is 3 eV, which corresponds to light energy in the near UV. When incorporated in paint, it serves to absorb UV light and minimizes paint discoloration due to sun exposure. High-end cosmetics contain nanoscale titanium dioxide particles that minimize skin damage by absorbing UV light in the solar spectrum.

oxygen vacancies are produced. What is the carrier type? (Hint: Since electrical neutrality has to be maintained, each oxygen vacancy still contains two electrons. Are these two electrons more or less tightly bound with the absence of oxygen?)

3.8 SELECTED SEMICONDUCTOR DEVICES

In this section, we describe the principles and operations of a few selected semiconductor devices: the Hall probe, p–n junction (with its various incarnations), bipolar junction, and field effect transistors.

3.8.1 HALL PROBE

A Hall probe is a device to measure magnetic field strength. Consider a piece of semiconductor oriented relative to a magnetic field B as shown in Figure 3.8. A current I is passed through the semiconductor. Assume that the majority charge carriers are holes. The magnetic field deflects the holes towards one side of the semiconductor as shown. This is known as the Lorentz force, given by $e(v \times B)$, where v is the carrier (hole) velocity. The accumulation of positive carriers on one side continues until the electrostatic repulsive force due to this charge accumulation equals the Lorentz force:

$$e(v \times B) = eE_H,\qquad(3.8)$$

where E_H is known as the Hall field (the Hall voltage V_H divided by the width of the semiconductor). By expressing the carrier velocity in terms of the current I, we obtain:

$$V_H = \frac{BI}{t}\frac{1}{pe},\qquad(3.9)$$

where t is the semiconductor probe thickness and p the hole concentration.

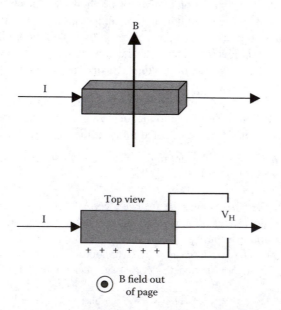

FIGURE 3.8 The Hall effect.

Therefore, measurement of the Hall voltage directly gives the magnetic field strength when the other quantities (current, probe thickness, and carrier concentration) are known. Alternatively, if the magnetic field is known, we can use this technique to determine semiconductor carrier concentration.

EXAMPLE

To get a feel of the signal involved, consider a Hall sensor with the following specifications: sensor thickness = 0.1 mm; hole concentration = $1 \times 10^{17}/cm^3$. At a sensor current of 5 mA, express the sensitivity of the Hall probe in millivolts per tesla.

Solution

Using the notation of Equation (3.9), $t = 10^{-4}$ m, $p = 1 \times 10^{23}/m^3$, and $I = 5 \times 10^{-3}$ A. The Hall voltage per unit tesla V_H is given by:

$$V_H = \frac{5 \times 10^{-3}}{1 \times 10^{23} \times 1.6 \times 10^{-19} \times 10^{-4}}$$

$$= 3.1 \times 10^{-3} \text{ V}.$$

Therefore, the sensitivity of the Hall probe is 3.1 mV/T.

Questions for discussion:

1. Can we determine the semiconductor carrier type (i.e., n- or p-type) using Hall measurements? (Hint: Study the deflection of positive and negative charge carriers in Figure 3.8, remembering that conventional current corresponds to the flow of positive charges in the same direction and flow of negative charges in the opposite direction.)
2. Semiconductors are better materials than metals for use in Hall probes to measure magnetic field strengths. Why? (Hint: Examine Equation 3.9. What is the key material parameter in that equation?)

3.8.2   P–N JUNCTION

The semiconductor p–n junction is one of the most basic semiconductor devices. p–n junctions are used as rectifiers (converting alternating currents into direct currents), light-emitting diodes, lasers, x-ray detectors, and solar cells, as well as integral components in transistors and storage devices.

As the name implies, a p–n junction consists of a p-type semiconductor grown on top of an n-type semiconductor.* Recall that the majority charge carriers are electrons in n-type and holes in p-type semiconductors. When these two materials are placed next to each other, this difference in charge carrier concentration results in diffusion across the junction (Figure 3.9). The diffusion of holes from the p-side to the n-side (and electrons from n to p) continues until the opposing electric field is strong enough to stop this diffusion. This electric field causes the conduction and valence bands to shift as shown in Figure 3.9.

The band bending shown in Figure 3.9 is the critical feature of this device. In any given instance, there are currents flowing across the junction. Consider electrical currents originating from the n-side:

---

* Placing a piece of p-Si mechanically on top of a piece of n-Si will not produce the effects described here. This is due to the presence of an insulating native oxide (SiO_2) that blocks charge flow.

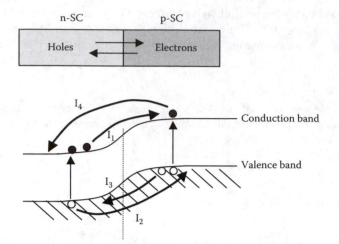

FIGURE 3.9 Energy band diagram and current flow for a *p–n* junction at thermal equilibrium.

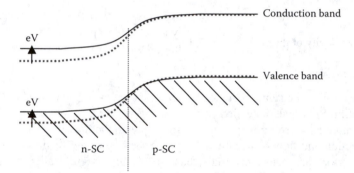

FIGURE 3.10 Energy band diagram of a *p–n* junction when a negative voltage bias is applied to the *n*-side. Note the reduction of the amount of band bending.

- Electrons in the conduction band may be thermally activated, climbing over the electrostatic barrier to go to the *p*-side. These electrons may then recombine with holes on the *p*-side. Let us label this electron current as I_1.
- Thermal excitation of valence electrons across the bandgap can occur, with the holes "floating" over the *p*-side. Let us label this hole current as I_2.

Therefore, the net positive current flowing from the *n*- to the *p*-side is $-I_1 + I_2$. Likewise, we will have currents due to holes thermally activated from the *p*- to the *n*-side (I_3) and downhill flow of thermally excited electrons in the conduction band (I_4). The net positive current flowing from the *p*- to the *n*-side is then $I_3 - I_4$. At thermal equilibrium, the net current must be zero:

$$-I_1 + I_2 = I_3 - I_4. \tag{3.10}$$

Consider what happens when we apply a small negative bias V to the *n*-side. This raises the conduction and valence band positions of the *n*-side by the same amount (eV) relative to the *p*-side, as shown in Figure 3.10.

Let us examine what happens to each of these four terms. Since the barrier to climb is reduced by eV, we expect I_1 to increase by the Boltzmann factor $\exp(eV/k_BT)$. Since I_2 and I_4 are due to

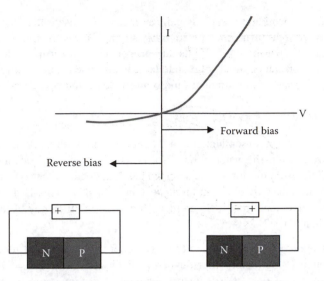

FIGURE 3.11 The current–voltage relationship for a p–n junction and the corresponding electrical connections.

thermal excitation across the bandgap, we do not expect these two currents to be affected by external voltage bias. Likewise, we expect I_3 to increase by a factor of $\exp(eV/k_BT)$. Therefore, we have the following situation after applying a negative voltage V on the n-semiconductor:

$I_1 \rightarrow I_1 \exp(eV/k_BT)$
$I_2 \rightarrow I_2$
$I_3 \rightarrow I_3 \exp(eV/k_BT)$
$I_4 \rightarrow I_4$

It is left as an exercise to show that the net current I flowing through the p–n junction when a voltage V is applied is given by:

$$I = -(I_1 + I_3)\left[\exp\left(\frac{eV}{k_BT}\right) - 1\right]$$

$$= I_o\left[\exp\left(\frac{eV}{k_BT}\right) - 1\right]. \tag{3.11}$$

This current–voltage relationship is shown in Figure 3.11. The asymmetric and nonlinear dependence of current on applied voltage is sometimes called nonohmic behavior. This property of the p–n junction is used in rectifiers to convert alternating currents (AC) into direct currents (DC).

3.8.3 LIGHT-EMITTING DIODES AND LASERS

When we pass a current through a p–n junction, we are injecting minority charge carriers, which eventually recombine with carriers of the opposite sign. If silicon is used to make a p–n junction (or diode, as it is sometimes called), the recombination of electrons and holes gives only heat.*

---

* It is beyond the scope of this text to explain why it is so. It suffices to say that the electronic structure of silicon is such that there is a large difference between the momentum of electrons in the conduction band and holes in the valence band. Since photons have negligible momentum, the large momentum difference between electrons and holes makes it difficult to conserve energy and momentum simultaneously in the light emission process for silicon.

However, if another semiconductor such as gallium arsenide is used, the electrical energy will be efficiently converted to light (efficiency of 30–60% is typical). The resulting device is known as a light-emitting diode. With sufficient currents and proper geometry to provide an optical cavity in which light amplification can occur, a solid-state laser is produced. These solid-state lasers are now widely used in data storage, entertainment (audio and video), and optical communications.*

3.8.4 SOLAR CELLS AND X-RAY DETECTORS

As mentioned earlier, when a semiconductor is illuminated with photons of energy $h\nu \geq E_g$, excitation of electrons from the valence band to the conduction band occurs. If this is to occur within the region of the p–n junction where the band bending occurs, electrons move in one direction and holes move in the other, resulting in an electric current flow. This is the basis of a solar cell converting light into electrical energy. Solar energy conversion efficiency of the best single-crystal silicon solar cells is about 22–25%.**

Question for discussion: Gallium arsenide (bandgap energy = 1.35 eV at room temperature) provides better solar energy conversion efficiency than silicon does (bandgap = 1.1 eV). This does not seem to make sense since there is less energy available for conversion with the larger bandgap material.*** Explain this apparent paradox. (Hint: What is the relationship between the maximum voltage output from the solar cell and the bandgap of the semiconductor?)

With slight adaptations, the same p–n junction can be used to detect and the measure energy of x-rays. When x-rays are incident on a given material, energetic electrons from the material are ejected. This is known as the photoelectric effect.**** When this occurs in a semiconductor such as silicon, the energetic electron (the photoelectron) can excite valence electrons to the conduction band, thus producing electron–hole pairs. Based on the electronic structure of silicon, the average energy to produce one electron–hole pair is 4.4 eV. Therefore, one 4.4-keV x-ray photon incident on silicon produces, on average, 1,000 electron–hole pairs. With proper electronics, this signal can be detected and measured precisely. This is known as an energy-dispersive x-ray (EDX) detector. In combination with an electron microscope, this detector is used routinely in quantitative analysis of materials from the macro- to the microscopic scale.

Question for discussion: In the preceding discussion, one 4.4-keV x-ray photon produces on average 1,000 electron–hole pairs. The electron–hole pairs are separated and detected. How can a charge as small as 1,000 electrons be detected? (Hint: The trick is to collect the charge and produce a voltage. Once a voltage is produced, modern electronics can do wonders. What is a common circuit component that develops a voltage when you place a charge on it?)

---

* Depending on applications, the wavelength of light used is different. For example, in communications using optical fibers, light with wavelength around 1.3 μm is used. At this wavelength, light absorption by the optical fiber is minimum. Fewer repeaters are needed to amplify the signal. For optical data storage, one would like to use shorter wavelengths to write smaller bits, resulting in larger data capacity.

** One of the challenges of making practical solar-powered cars is the low solar energy flux. With the Sun directly overhead, the solar energy flux on the Earth's surface is about 1400 W/m². With 20% conversion energy, 10 m² solar cells produce 2800 W. If this flux is sustained for 4 h and the energy is stored, the total amount of energy can run a small car for about 30 miles.

*** It is useful to note that solar power peaks at a wavelength of 460 nm (green). Nature understands that very well with the proliferation of green vegetation. It is interesting to speculate how plant life would evolve if the sun became hotter over time.

**** Einstein provided the explanation for the photoelectric effect, for which he was awarded the Nobel Prize in the early 1900s. Back then, although there was a desire to award Einstein this prize based on his work on special relativity, it was rumored that very few people on the Nobel committee really understood special relativity. Instead, they gave him the prize based on something they understood: explanation of the photoelectric effect.

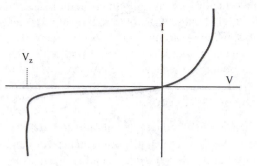

FIGURE 3.12 Current–voltage characteristics of a Zener diode.

3.8.5 VOLTAGE REGULATOR (ZENER DIODE)

In a regular diode under reverse bias, the current is small. As the reverse bias increases, the electric field increases. Conduction electrons gain more and more energy from this electric field between collisions with valence electrons. At some critical reverse bias voltage, the collisional energy transfer is sufficient to excite valence electrons across the bandgap. This increases the conduction electron concentration. This process repeats, thus producing even more conduction electrons and hence a large current. This current–voltage curve shows a rapid increase of current beyond a certain reverse-bias voltage as shown (Figure 3.12). This is known as Zener breakdown. This type of diode is used as a voltage regulator. At the designed voltage V_z, the output voltage is independent of current.

3.8.6 BIPOLAR JUNCTION TRANSISTOR

There are two types of bipolar junction transistors: NPN and PNP. We will discuss the NPN transistor here. A transistor is used mostly for amplifier and switch applications. As the name suggests, an NPN transistor consists of three parts:

- Emitter, a heavily doped n-type semiconductor
- Base, a thin and moderately doped p-type semiconductor
- Collector, a lightly doped n-type semiconductor

When set up as an amplifier, the emitter–base junction is forward biased, while the base–collector junction is reverse biased (Figure 3.13). In this configuration, electrons are injected from the emitter to the base. The base is designed to be so thin that most of the conduction electrons drift directly into the collector. In typical transistors, 99% of the current injected into the base goes to the collector. Another way to look at it is that the ratio of collector current to base current is about 100. Now imagine what happens when a small voltage is applied to the base. This voltage modulates the height of the barrier that conduction electrons have to climb in order to get into the base (and hence into the collector). Consequently, the collector current I_C is very sensitive to the voltage at the base-emitter junction V_{BE} applied to the base.

Assume that the collector current flows through a resistance R_C. The voltage change across this resistance ΔV_C, equal to $(\Delta I_C)R_C$, depends sensitively on V_{BE}. The voltage gain, $\Delta V_C/\Delta V_{BE}$, in typical circuits ranges between 100 and 500.*

As suggested by Figure 3.13, the transistor can also function as a switch. The base–emitter bias V_{BE} controls the flow of electrical current from the emitter to the collector. In normal operations,

---

* Note that the voltage gain is due to the transfer of current flow from a low-resistance circuit (base–emitter junction) to a high-resistance circuit (base–collector junction). This is the origin of the name "transistor" (transfer of resistance).

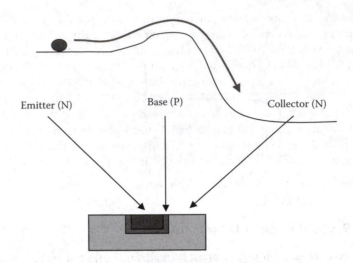

FIGURE 3.13 Motion of electrons in the conduction band of an NPN transistor and the individual components (emitter, base, and collector).

the base–emitter junction is forward biased, and V_{BE} is typically about 0.6 eV for silicon-based transistors (positive on base). When V_{BE} is set to zero or negative, little or no current flows from emitter to collector.

3.8.7 FIELD EFFECT TRANSISTOR

There are different types of field effect transistors (FET). Here, we will discuss the operation of an *n*-channel metal-oxide semiconductor field effect transistor (MOSFET), as shown in Figure 3.14(a). The starting point of such a device is a lightly doped *p*-type semiconductor. Voltage is applied between two points, known as source and drain. The source and drain regions are heavily doped with *n*-type impurities (*n*+). A voltage V_{GS}, applied on a metal contact (known as the gate) through an insulating oxide between the source and the drain, controls the drain current I_D.

When the gate voltage is zero, electrical conduction in the channel between the source and the drain is normal. When a negative voltage is applied to the gate, the resulting electric field repels conduction electrons in the channel and hence creates a region with depleted conduction electron concentration (depletion region). This reduces the drain current as shown in Figure 3.14(b) and thus provides the required transistor action.

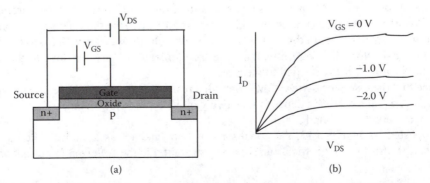

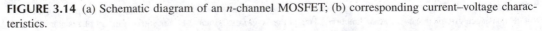

FIGURE 3.14 (a) Schematic diagram of an *n*-channel MOSFET; (b) corresponding current–voltage characteristics.

FETs differ in two ways from bipolar junction transistors discussed in the preceding section. First, since the control signal is applied to the gate, which is electrically isolated from the n-channel, the input impedance is very high (1000 MΩ or higher). If care is not taken, electrostatic charging can occur, resulting in high voltage buildup and destruction of the transistor.* In comparison, bipolar junction transistors normally operate with forward-biased input circuits with relatively low input impedance (a few kilohms). Second, the FET has much lower noise than the bipolar junction transistor. Any noise in the input base current in the bipolar junction transistor is amplified (typically ~100 times) in the collector circuit. In the MOSFET, there is little or no gate current. The only input noise comes from thermal (Johnson) noise** across a resistor in the input circuit. With such low-noise characteristics, the FET is often used in the first stage of a low-noise preamplifier.

3.9 ELECTRON TUNNELING

Every time we switch on a light, we experience the benefits of electron tunneling. An oxide layer is always present on the surface of an electrical conductor, such as copper or aluminum, exposed to air. In bulk form, these oxides are insulating. However, when the thickness decreases to the 1- to 5-nm range, these normally insulating oxides do not seem to offer much electrical resistance at all. Electrons act as if they can punch or tunnel through the thin insulating layer.

One way to understand this electron tunneling phenomenon is to use the uncertainty principle. This principle states that in the atomic world, energy cannot be measured precisely. If it takes a certain amount of time Δt to measure the energy of an electron, then the uncertainty in the electron energy ΔE will be such that $\Delta E \Delta t \approx \hbar/2$, where $\hbar$ is Planck's constant divided by 2π. Let us assume that the insulating layer presents a barrier to the conduction electrons, height $V > E$, as shown in Figure 3.15.

Without thermal excitation, conduction electrons incident on this insulating barrier will be reflected — there will be no electrical conduction across this barrier. However, if the energy uncertainty ΔE is equal to $V - E$, then the electron will have just enough energy to go over the barrier. This is only possible if the time Δt allowed to measure the electron energy is equal to $\hbar/(2\Delta E)$. During this time, the electron could have traveled through the energy barrier if the barrier was thin enough. We can almost think of this scenario as the conduction electron, in a sleight of hand, borrowing energy $V - E$ for a small enough amount of time without anyone knowing about it, at least without violating the uncertainty principle. Quantum mechanics shows that in order for

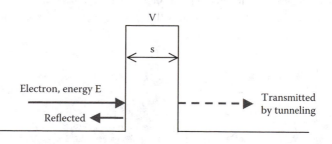

FIGURE 3.15 Electron tunneling through a barrier.

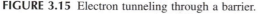

* Many add-on devices for computers contain MOSFETs. Because of their sensitivity towards static electricity, they are shipped inside conducting plastic bags. Before handling such add-on devices, one must get rid of static charge by touching ground (e.g., water pipe or faucet).

** Thermal fluctuations give rise to noise voltages. The root-mean-square noise voltage V_{rms} appearing in a resistance R is given by $V_{rms} = \sqrt{4k_B TR\Delta f}$, where k_B is the Boltzmann constant, T absolute temperature, and Δf the frequency bandwidth. For example, at 300 K with $R = 1$ MΩ and $\Delta f = 1$ MHz, the r.m.s. noise voltage ≈ 0.13 mV.

appreciable electron tunneling to occur through an insulator, the insulator thickness s should be on the order of:

$$4\sqrt{\dfrac{\hbar^2}{2m(V-E)}}\,,$$

where m is the electron mass.

EXAMPLE

Let us assume that the electron energy is 2 eV below the top of the barrier. Estimate how thin the barrier should be in order to obtain appreciable tunneling ($m = 9.1 \times 10^{-31}$ kg, $\hbar = 1.05 \times 10^{-34}$ J-s, and 1 eV = 1.6×10^{-19} J).

Solution

In this problem, $V - E = 2$ eV. For appreciable tunneling, the insulator thickness s required is given by:

$$s \simeq 4\sqrt{\dfrac{\hbar^2}{2m(V-E)}}$$

$$= 4\sqrt{\dfrac{(1.05\times10^{-34})^2}{2\times9.1\times10^{-31}\times2\times1.6\times10^{-19}}}$$

$$= 5.5\times10^{-10}\,m.$$

In terms of impact to our daily lives, field-emission displays are potentially the most common application of tunneling. Each pixel of a field-emission display consists of a sharp tip of radius R to which a negative voltage, $-V$, is applied. The electric field at the tip surface is $\approx V/R$. Setting $R = 100$ nm and $V = 40$ V yields an electric field of 4×10^8 V/m. This electric field decreases the barrier height at the rate of 1 eV every 2.5 nm, as shown in Figure 3.16. For most materials, the barrier that confines the electrons has a typical height of 4–5 eV. Therefore, the electric field reduces the

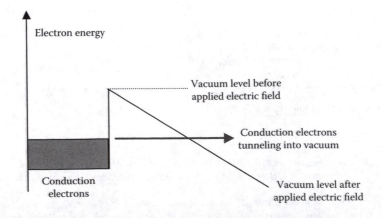

FIGURE 3.16 Electron energy diagram near the surface of a field-emission tip.

width of the tunneling barrier (to about 10–12 nm)* so that electrons from the tip can tunnel into vacuum without the need for thermal excitation. This is known as field emission. The field-emitted electrons then impinge on a phosphor and light up the screen.

Because of the ease with which electrons can be extracted from a sharp tip by tunneling, we use field-emission tips as an alternative to traditional hot filaments to generate electron beams in electron microscopes. In 1981, the scanning tunneling microscope (STM) was invented by Gerd Binnig and Heinrich Rohrer at IBM Zurich. Based on electron tunneling from a sharp tip to a sample surface (spaced at about 1 nm from the tip), Binnig and Rohrer developed a tool that can image surfaces with atomic resolution (in air, liquids, and vacuum) and manipulate atoms. Many new and exciting tools and devices have evolved from the basic STM. For this invention, they were awarded the Nobel Prize in physics in 1986.

Question for discussion: The scanning tunneling microscope works with a sharp tip, local radius probably ~1 nm, located at ~1 nm from the specimen surface. It is understandable why it works in vacuum, analogous to the field emission microscope. In air or liquids, however, molecules from the ambient can scatter electrons — yet, the STM has no problem operating under these environments. Explain.

3.10 THIN FILMS AND SIZE EFFECTS

Our earlier discussions show that the electrical conductivity σ can be written as $ne^2\tau/m$, where τ is the average time between collisions. Writing $\tau = \lambda/v$, where λ is the average distance between collisions (mean free path) and v is the average speed of conduction electrons, we can write:

$$\sigma = \frac{ne^2\lambda}{mv}. \tag{3.12}$$

For most bulk metallic conductors, the mean free path ranges between 10 and 100 nm at room temperature. Consider what happens when the conductor is in the form of a thin film, with thickness approaching the mean free path. Not only will electrons lose energy by colliding with phonons, impurities and other electrons, but some will also do so by colliding with the top and bottom surfaces of the thin film. Since the average grain size of most polycrystalline thin films is the same order of magnitude as the film thickness, a substantial fraction of conduction electrons may also lose energy via scattering by grain boundaries. Therefore, the electrical conductivity of thin films decreases with decreasing film thickness. This is known as the classical size effect.** All properties that depend on electron transport — for example, thermal conductivity, Hall effect, thermoelectric effect, etc. — exhibit the classical size effect.

Question for discussion: Equation (3.9) indicates that the Hall voltage increases with decreasing thickness t of the semiconductor probe. Is there a limit to this trend?

---

* The width of the tunneling barrier is equal to the barrier height φ divided by the eE, where e is the electron charge and E the electric field.

** For those who are interested, the electrical conductivity of single-crystal thin films (i.e., no grain boundary scattering) is given by the following formula:

$$\frac{\sigma_f}{\sigma_o} = 1 - \frac{3}{2\kappa}(1-p)\int_1^\infty \left(\frac{1}{t^3} - \frac{1}{t^5}\right)\frac{1-e^{-\kappa t}}{1-pe^{-\kappa t}}\,dt,$$

where σ_f is the film conductivity, σ_o the bulk conductivity, $\kappa = a/\lambda$, a = film thickness, λ = mean free path, and p is the fraction of electrons incident on the film surfaces that are specularly scattered.

When the film thickness approaches the de Broglie wavelength of conduction electrons,* the discrete nature of electronic energy levels becomes apparent. Electron transport properties such as electrical conductivity exhibit oscillations as a function of film thickness. This is known as the quantum size effect. For most materials systems, quantum effects become important when the system size is in the 1- to 10-nm regime or below.

Why should quantum effects concern us? Some have been exploited to produce extraordinarily sensitive sensors of magnetic fields. In fact, these devices are so sensitive that they can detect magnetic field changes in our brains when we think different thoughts (mind readers?). Infrared detectors based on quantum effects can detect the lighting of a match from many miles away. Cryptographers want to use quantum effects (based on the probabilistic nature of occupation of quantum states) in materials to develop unbreakable codes or encryption schemes. Many scientists suggest that quantum computers can run much faster, while others believe that new thermoelectric energy-conversion devices based on quantum effects can be developed with higher efficiency. There is much to discover and learn.

3.11 THERMOELECTRIC ENERGY CONVERSION

Connecting two different conducting materials A and B together (as shown in Figure 3.17) with the junctions held at temperatures T_{hot} and T_{cold} yields a voltage output V proportional to the temperature difference $\Delta T = (T_{hot} - T_{cold})$:

$$V = (S_A - S_B)\, \Delta T, \qquad\qquad (3.13)$$

where S_A and S_B are the Seebeck coefficients for materials A and B, respectively. This is the phenomenon we exploit to measure temperatures in thermocouples. For example, chromel–alumel

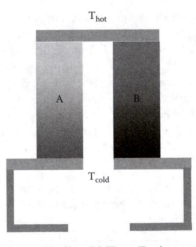

$$V = (S_A - S_B)\,(T_{hot} - T_{cold})$$

FIGURE 3.17 Illustration of the Seebeck or thermoelectric effect: V = voltage output; S_A = Seebeck coefficient of material A; S_B = Seebeck coefficient of material B; T_{hot} = hot junction temperature; T_{cold} = cold junction temperature.

---

* As a reminder, the de Broglie wavelength λ of a particle with mass m and velocity v is equal to $\lambda = h/mv$, where h is the Planck constant.

thermocouples (made of chromium–nickel and aluminum–nickel alloys) have an average thermo-electric response $(S_A - S_B)$ of ~40 µV/K near room temperature.

Because of the temperature difference, heat flows from the hot to the cold junction. Therefore, heat energy must be supplied to the hot junction to maintain the temperature difference. The Seebeck effect provides the solid-state approach to converting heat energy into electricity. As discussed later, with proper materials, this phenomenon may provide a partial solution to the global energy problem by converting waste heat into useful work.*

The electrical power output due to the Seebeck effect is equal to V^2/R, where R is the resistance of the material. This in turn is proportional to σS^2, where σ is the electrical conductivity and S the Seebeck coefficient of the material. At the same time, because of the temperature gradient, heat flows from the hot to cold junction. For a given geometry, the heat flux is proportional to the thermal conductivity K. Therefore, the ratio of electrical power output to heat flux input should scale as $\sigma S^2/K$, which is considered the figure of merit, Z, for thermoelectric power conversion. The dimension of Z is inverse temperature. We can define a dimensionless thermoelectric figure of merit ZT:

$$ZT = \frac{\sigma S^2 T}{K}.$$ (3.14)

The energy conversion efficiency η for such a device is given by:

$$\eta = \frac{\Delta T}{T_{hot}} \frac{(M-1)}{M + \dfrac{T_{cold}}{T_{hot}}},$$ (3.15)

where $M = (1 + ZT_{hot})^{1/2}$.

In the limit $M \gg 1$, the energy conversion efficiency approaches $\Delta T/T_{hot}$. The latter quantity is known as the Carnot efficiency: the maximum efficiency to be attained in converting heat into useful work. Figure 3.18 plots the energy conversion efficiency versus the temperature difference ΔT for various figures of merit, ZT, at a cold junction temperature of 400 K. For example, at $T_{cold} = 400$ K and $\Delta T = 400$ K, we can obtain efficiencies in the range of 25% for $ZT = 5$.** Doped semiconductors containing tellurium (e.g., PbTe, Bi_2Te_3, and Sb_2Te_3) have the optimum combination of electrical conductivity, Seebeck coefficient, and thermal conductivity to give ZT values in the 2–5 range. Work on certain Si/SiGe nanolayer structures (known as quantum wells) suggests new opportunities to increase ZT values even further.

---

* The reverse phenomenon also occurs — that is, passing an electrical current causes one junction to become hot and the other junction to cool. This is known as the Peltier effect. This phenomenon has been exploited in small refrigerators and integral car seat coolers.

** By definition, a thermoelectric energy conversion device consists of two materials. The figure of merit must involve physical properties of both materials, as given by:

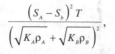

where ρ represents the electrical resistivities of the two components.

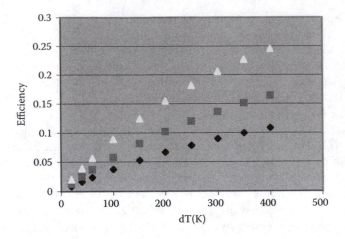

FIGURE 3.18 Energy conversion efficiency of a thermoelectric generator versus temperature difference between the hot and cold junctions.

3.12 ELECTRICAL SIGNALING IN NEURONS: LESSONS FROM MOTHER NATURE

The transmission of nerve impulses from one nerve cell to another is different from electrical conduction in metals and semiconductors. Figure 3.19 is a schematic diagram of a nerve cell (neuron). One major characteristic of the neuron is the axon, which can be more than a meter long. In the resting state, there is a higher concentration of potassium ions (K^+) than sodium ions (Na^+) inside the axon, while the opposite is true just outside the axon. Inside mammalian cells, K^+ and Na^+ concentrations are 140 and 5–15 mM, respectively. Outside mammalian cells, K^+ and Na^+ concentrations are 5 and 145 mM, respectively. Potassium–sodium ion exchange pumps (proteins) on the surface of the cell membrane maintain this concentration difference. As a result, the protein inside the axon is negatively charged, and there is a 60 mV potential difference across the cell membrane.*

When a neuron is stimulated at a given location, the location is suddenly "depolarized" (Na^+ channels open), while K^+ ion channels close, resulting in a sudden influx of Na^+ into the cell.** The

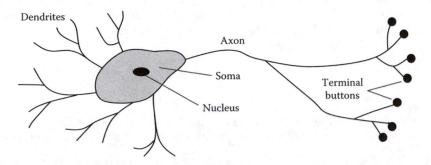

FIGURE 3.19 Schematic diagram of a typical neuron.

---

* The potential ΔV due to ion concentration difference is given by ($k_B T/ne$) ln R, where k_B is the Boltzmann constant, T absolute temperature, n charge on the ion, e electron charge, and R concentration ratio.

** As discussed in Chapter 2, ion channels are specific sites on the cell membrane that are triggered to open or close by signaling molecules or electrical potentials.

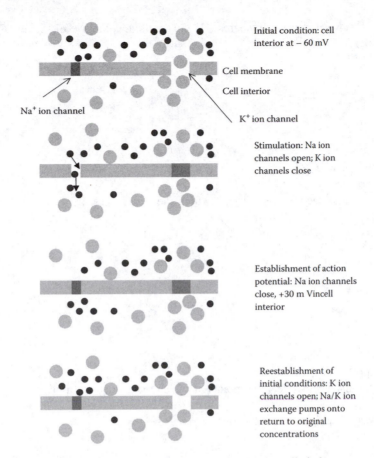

Initial condition: cell
interior at − 60 mV

Cell membrane

Cell interior

Na⁺ ion channel

K⁺ ion channel

Stimulation: Na ion
channels open; K ion
channels close

Establishment of action
potential: Na ion channels
close, +30 m Vincell
interior

Reestablishment of
initial conditions: K ion
channels open; Na/K ion
exchange pumps onto
return to original
concentrations

FIGURE 3.20 Propagation of a nerve impulse. K⁺, large circles; Na⁺, small circles.

primary driving force for sodium ion migration is the concentration gradient. In about 1 ms, the membrane potential changes from –60 to +30 mV. This sudden shift of ions produces the neural impulse or action potential. Then, the sodium ion channels close and the potassium ion channels open. Sodium–potassium ion exchange pumps on the cell membrane gradually reestablish the normal sodium and potassium ion concentrations within the cell, causing the membrane potential to return to the resting state of –60 mV (Figure 3.20).

The transmission of a nerve impulse is via the propagation of this depolarization wave along the axon. When the action potential arrives at the end of the axon, certain chemicals known as neurotransmitters are released and received by an adjacent neuron. These neurotransmitters bind to certain receptor sites and control the opening and closing of ion channels, thus continuing the propagation of the nerve impulse. Unlike normal electrical signal transmission, nerve impulses act like dominos and are transmitted from one neuron to another without attenuation.

APPENDIX: OHM'S LAW AND DEFINITIONS

Ohm's law defines the relationship between electrical current and applied voltage:

$$I = \frac{V}{R},$$

(3.16)

where I is the electrical current (amperes), V the applied voltage (volts), and R the electrical resistance (ohms).

DEFINITIONS

$$J = \frac{I}{A},$$

(3.17)

where

J = the *current density*
I = the current
A = the cross-section area through which the current passes.

The unit for current density is amp/m².

*   *   *

$$E = \frac{V}{L},$$

(3.18)

where

E = the *electric field*
V = applied voltage
L = the distance between the points at which the voltage is applied.

The unit for electric field is volts per meter.

*   *   *

An alternative statement of Ohm's law is

$$J = \sigma E,$$

(3.19)

where

J = the current density
σ = electrical conductivity
E = electric field.

The unit for electrical conductivity is (ohm-m)$^{-1}$.

*   *   *

$$R = \frac{\rho L}{A},$$

(3.20)

where

R = the resistance
ρ = electrical resistivity ($1/\sigma$)
L = distance between the points of voltage application
A = the cross-section area.

The unit for resistance is ohm.

$$*\quad*\quad*$$

$$g = \frac{1}{R},\tag{3.21}$$

where g is the conductance. The unit for conductance is ohm^{-1} or siemen (S). Electrical conductivity is also expressed as S/m.

PROBLEMS

1. The electrical conductivity of Ag is 6.3×10^7 (ohm-m)$^{-1}$ at room temperature. There are 5.85×10^{22} silver atoms per cubic centimeter. Assuming one conduction electron per silver atom,
   a. Calculate the electron mobility in silver.
   b. Deduce the relaxation or mean collision time.
   c. Calculate the drift velocity of electrons in silver at an applied electric field of 1 V/cm.
2. A pure gallium arsenide (GaAs) single crystal (a III–V semiconductor) is doped with 1×10^{17} sulfur atoms/cm^3. Sulfur is a group VI element that goes into As sites.
   a. Is this a p- or n-type semiconductor?
   b. Assuming complete ionization of sulfur atoms at room temperature, what is the majority carrier concentration?
   c. Above room temperature, the conduction electron concentration increases rapidly with temperature. Explain the source of these additional carriers.
3. At 300 K or above, the electrical conductivity σ of silicon is proportional to exp $(-E_g/2kT)$, where E_g is the bandgap of silicon (1.1 eV). Given that it is possible to measure conductivity change $(d\sigma/\sigma)$ of 0.01%, show that, at around 300 K, silicon as a thermistor can detect temperature changes as small as 0.0016 K (1.6 mK). Under the same conditions, metals have a temperature coefficient of resistance $(1/R\ dR/dT) \sim 4 \times 10^{-3}$/K. Show that metal thermistors are at least 10 times less sensitive than silicon thermistors. (Hint: Write $\sigma = $ constant $\times$ exp $(-E_g/2kT)$. Determine the relationship between $d\sigma$ and dT.)
4. Sketch and compare the energy band diagrams when a PN junction is (a) forward biased; and (b) reverse biased with one that is unbiased.
5. In a light-emitting diode, what is the relationship between the emitted light energy and the bandgap of the semiconductor?
6. In the text, it is stated that $np = $ constant when donors or acceptors are introduced to a semiconductor. Would this still be true when donors and acceptors are introduced at the same time? Justify with a short explanation.
7. Starting with pure Si, we introduce 1×10^{16} boron atoms/cm^3 and 8×10^{16} phosphorus atoms/cm^3. What type (n or p) of semiconductor do we have? Justify with a short explanation.
8. For pure silicon at room temperature, the concentration of electrons in the conduction band is equal to that of holes in the valence band (1.45×10^{16}/m^3). Now we introduce phosphorus impurities (phosphorus is a group V element) at a concentration of 1×10^{21}/m^3 into silicon, making it an n-type semiconductor. Assume that all dopants are ionized and $e = 1.6 \times 10^{-19}$ C.
   a. What is the conduction electron concentration?
   b. What is the hole concentration?
   c. Given that the electron mobility is 0.135 m^2/V-s and hole mobility is 0.048 m^2/V-s, what is its electrical conductivity in (ohm-m)$^{-1}$?

   d. Now we introduce small amounts of boron (less than $1 \times 10^{21}/m^3$) into this *n*-type semiconductor (boron is a group III element). The electrical conductivity is found to decrease. Explain.

9. Look at the following section of the periodic table:

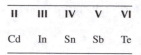

II	III	IV	V	VI
Cd	In	Sn	Sb	Te

   a. Consider the compound semiconductor CdTe. If Sn impurities are introduced into CdTe and sit on Cd sites, what semiconductor type would result (*p* or *n* type)? Explain.

   b. Consider the compound semiconductor InSb. In growing this semiconductor, excess indium atoms are introduced. These excess indium atoms sit on Sb sites. Explain which semiconductor type would result.

10. At room temperature, a small fraction of valence electrons gain enough thermal energy to be excited across the bandgap of pure silicon (1.1 eV).

   a. Sketch the spatial distribution of electrons in the valence band as referenced to Si atom positions. (Hint: Valence electrons are responsible for bonding in Si.)

   b. Sketch the spatial distribution of conduction electrons.

   c. We now expose this silicon sample to light and measure its electrical conductivity. The variation of electrical conductivity σ versus incident light energy $h\nu$ is shown in Figure 3.21. Explain the sudden increase of electrical conductivity above a certain light energy *x*. What is the approximate value of *x* in eV?

11. Zinc oxide ($Zn^{2+} O^{2-}$) is a II–VI semiconductor, which is a potential material for fabricating UV lasers.

   a. Oxygen vacancies are present in a zinc oxide crystal at 1 ppm concentration. The carrier concentration is above the intrinsic value expected of stoichiometric zinc oxide. Do these oxygen vacancies act as donors or acceptors? Explain. (Hint: The material must remain neutral when oxygen is removed.)

   b. To attempt returning the carrier concentration to the intrinsic value, we introduce 1 ppm sulfur impurities into the preceding zinc oxide crystal. Sulfur and oxygen are in the same column of the periodic table. The carrier concentration decreases, but does not quite get back to its original value. Speculate what may have happened. The following information may be useful: radius of $O^{2-} = 0.14$ nm, radius of $S^{2-} = 0.184$ nm.

12. Consider titanium dioxide, which is a wide bandgap semiconductor. The valence of titanium is +4. Consider the introduction of scandium impurities with valence +3. Scandium atoms go into titanium sites. Discuss the conductivity type of this doped semiconductor.

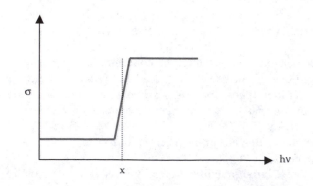

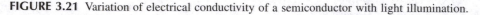

FIGURE 3.21 Variation of electrical conductivity of a semiconductor with light illumination.

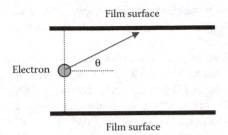

FIGURE 3.22 Thomson model to explain the classical size effect of electrical conductivity.

13. Consider stoichiometric In_2O_3. In this compound, indium has a valence of 3. Consider the doping of indium oxide by tin as a substitutional impurity, which has a valence of four. Discuss the conductivity type of this doped oxide. Indium tin oxide ($In_{2-x}Sn_xO_3$) belongs to a class of materials known as transparent conducting oxides widely used as displays, solar cells, and deicing applications.

14. Let us use a simple model to derive an equation relating electrical conductivity and film thickness. In bulk solids, the electrical resistivity is equal to A/λ_o, where A is some constant and λ_o is the mean free path due to intrinsic collision processes. In single-crystal thin films (no grain boundaries), the electrical resistivity is equal to $A(1/\lambda_o + 1/\lambda_t)$, where λ_t is the average distance between electron collisions with film surfaces. Consider a conduction electron located exactly in the middle of a thin film of thickness t as shown in Figure 3.22. Show that $1/\lambda_t = 4/(\pi t)$.

    Comment: This was the original theory advanced by Thomson in 1901 to explain the classical size effect. If we were to repeat this calculation by assuming an arbitrary starting position along the film thickness direction for the conduction electron and computing an average $1/\lambda_t$, we would obtain a divergent result due to electrons starting from the top and bottom surfaces. This problem was subsequently solved by Fuchs in 1938 (see the section on thin films) and extended by Sondheimer to include magnetic field effects.

15. Consider a solar cell based on the PN junction. The current–voltage characteristics are defined by the standard equation:

$$I = I_o\left[\exp\left(eV/k_BT\right) - 1\right],$$

where the symbols have their usual meaning. In the presence of light with energy greater than the bandgap of the semiconductor, the preceding equation is modified by the addition of a term αG, where G is the solar intensity and α some positive constant, as follows:

$$I = I_o\left(\exp\left(eV/k_BT\right) - 1\right) - \alpha G.$$

a. Explain the minus sign in front of αG.

b. Derive the relationship between the voltage output of the solar cell and the solar intensity G. Show that at sufficiently high solar intensity under open-circuit conditions ($I = 0$),

$$V = \frac{k_BT}{e}\ln\left(\frac{\alpha G}{I_o}\right).$$

The preceding expression suggests that the output voltage will increase with light intensity. There must be a limit to the magnitude of the output voltage from a solar cell. Discuss what this limit is.

16. Consider a piece of silicon doped with donor and acceptor impurities, with concentration N_D and N_A, respectively. Recall that, at a given temperature, the relationship $n.p = $ constant always holds. In addition, not all donors and acceptors are ionized. Based on the principle of electrical neutrality, write down a relationship among the following four quantities: n, p, N_D^+ (ionized donors), and N_A^- (ionized acceptors).

17. Tunneling is sensitive to the thickness of the tunneling of the barrier. To illustrate, let us look at the tunneling of electrons through a barrier of height ΔV, thickness t. The probability of tunneling is given by:

$$\exp\left(-2\sqrt{\frac{2m\Delta V}{\hbar^2}}\,t\right),$$

where m is the electron mass and $\hbar = $ Planck's constant $h/(2\pi)$. For the purpose of this exercise, set $\Delta V = 4$ eV.

a. Show that the tunneling probability can be expressed as:

$$\exp[-20.5t(nm)].$$

b. Show that the electron tunneling current changes by almost a factor of 10 when the barrier thickness changes by 0.1 nm.

18. In the normal operation of an NPN transistor, the emitter–base junction is forward biased at a voltage of 0.6 V, while the collector–base junction is reverse biased. Consider the situation in which the base voltage is increased by 0.1 V relative to the emitter. Assume that $k_B T = 0.026$ eV.

a. Show that the collector current increases by a factor of about 50 times.

b. Does the preceding answer depend on the fraction of emitter current going into the collector?

4 Mechanical Properties

4.1 GOSSAMER CONDOR AND GOSSAMER ALBATROSS*

Since the dawn of civilization, we as a species have always wanted to fly like birds. We mastered the technique of powered flight only 100 years ago and have moved rapidly forward from slow flying biplanes, to turboprops, supersonic jets, and space flight; yet, the ability to fly like birds without engine power has eluded us for a long time.** Henry Kremer put up a prize for the first human-powered aircraft to complete a prescribed figure-eight course. Paul MacCready designed and built an aircraft named Gossamer Condor to do just that. He won the second Kremer prize with a similar aircraft named Gossamer Albatross by crossing the 22.25-mile British Channel on June 12, 1979, with human power alone (Figure 4.1). This flight took 2 h and 49 min and was powered solely by the pilot.

MacCready faced two problems in designing this aircraft: the amount of sustained human power available for flight and the strength-to-weight ratio of the aircraft. He recruited a well-conditioned bicyclist, Bryan Allen,*** as the pilot, who could produce about 3 W of sustained power per kilogram of weight, or about 200 W for his 145-lb ($\approx$65 kg) weight.**** To cross the British Channel in a reasonable amount of time (allowing for head winds) without driving the pilot to exhaustion, the airspeed should not be much less than 18 mph (8 m/s). At 200 W, the aircraft must exert a drag no greater than 25 N (200 W divided by 8 m/s) and at the same time provide sufficient lift to support the entire weight of the aircraft, including the pilot. This puts stringent constraints on the design of the aircraft. The final specifications are:

Span: 97.67 ft
Length: 34 ft
Wing area: 488 ft^2
Empty weight: 70 lb

To make a 70-lb aircraft strong enough to support itself and the pilot and to withstand forces due to occasional turbulence requires extraordinary materials. The skeleton of the Gossamer Albatross is made of carbon fiber tubing. Pound for pound, carbon fibers are about six times stronger than steel. The wing and control surfaces are covered with thin Mylar sheets, while the ribs and leading edges of the wings are made of polystyrene foam. The only metal parts of the aircraft are the pedals, cranks, drive chain, seatpost, and a few fittings and wires. Without the use of high-strength lightweight materials and an understanding of the mechanical behavior of materials and aerodynamics, the two Gossamer aircraft might not even have lifted off the ground.

---

* The U.S. aviation community uses British units. Both British and metric units are presented in this section.
** You can fly in a glider without engine power, but you need to have steady thermal currents to stay aloft.
*** Many things come full circle — the Wright brothers were in the bicycle business when they took their first flight on December 17, 1903.
**** It is instructive to estimate how much energy Bryan Allen expended in this flight. At an average power of 200 W for this 2.8-h flight, the total energy is equal to ~2×10^6 J. This is the caloric content contained in 48 g of fat, which is not much (about the amount in a cheeseburger). This suggests that, while exercise is an important part of any weight-reduction program, controlling caloric intake has to be an equally important consideration.

FIGURE 4.1 The Gossamer Albatross aircraft weighs only 70 lb when empty. It is made mostly of polymers and flies with an airspeed of about 18 mph. (The left figure is from NASA archives; the right figure is reprinted with permission from the Museum of Flight, Seattle, Washington.)

Aerodynamics Notes

How do we arrive at the final specifications of the Gossamer aircraft? The aerodynamic lift L is given by:

$$L = \frac{1}{2}C_L \rho v^2 A,$$

where
C_L = the lift coefficient
ρ = density of air
v = airspeed
A = the wing area

Setting $C_L = 0.55$, $\rho = 0.00238$ slug/ft³ (1 slug = 1 lb divided by 32), and $v = 26$ ft/s ($\approx$8 m/s), we can calculate the lift per unit wing area as:

$$\frac{L}{A} = \frac{1}{2}0.55 \times 0.00238 \times (26)^2 = 0.44 \text{ lb/sq.ft.}$$

For an aircraft with a total weight of 215 lb (70 lb of empty weight plus 145-lb pilot), we require a total wing area of:

$$A = \frac{215}{L/A} = \frac{215}{0.44} = 488 \text{ sq.ft.}$$

The other consideration is aerodynamic drag. The design objective is to obtain no more than 5.6 lb (25 N) of drag at a speed of 26 ft/s with 488 ft² of wing area (here, we ignore drag due to the fuselage). The aerodynamic drag D is given by:

$$D = \frac{1}{2}C_D \rho v^2 A,$$

where C_D is the drag coefficient. Substituting $D = 5.6$ lb, $\rho = 0.00238$ slug/ft^3, $v = 26$ ft/s, and $A = 488$ ft^2, we can calculate the drag coefficient as follows:

$$C_D = \frac{2D}{\rho v^2 A} = \frac{2 \times 5.6}{0.00238 \times (26)^2 \times 488}$$

$$= 0.014.$$

The drag coefficient C_D for thin wings with a few percent of relative camber (wing curvature) adopted in this aircraft is empirically given by:

$$C_D = 0.009 + \frac{C_L^2}{3(AR)},$$

where AR is the aspect ratio of the wing, defined as the length of the wing divided by the average width. Substituting $C_D = 0.014$ and $C_L = 0.55$, we can readily calculate the aspect ratio to be 20.

For a total wing area of 488 ft^2, the wing should measure about 98.8 ft by 4.9 ft for rectangular planform wings.

4.2 DEFINITIONS AND UNITS

The following definitions and units are commonly used in describing mechanical properties of materials.

4.2.1 Stress, Strain, and Young's Modulus

Consider a rectangular bar of initial length L_o and initial cross section A_o with force F applied as shown in Figure 4.2. Stress σ is defined as force F per unit area A_o:

$$\sigma = \frac{F}{A_o}. \tag{4.1a}$$

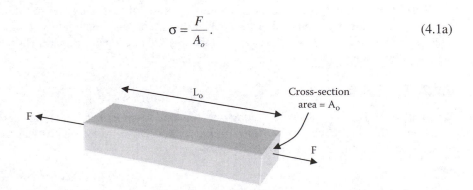

FIGURE 4.2 Specimen test geometry in tensile deformation.

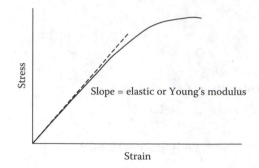

FIGURE 4.3 A typical stress–strain curve.

This definition is sometimes known specifically as engineering stress, to distinguish it from the definition of true stress, σ_T, which is defined as follows:

$$\sigma_T = \frac{F}{A}, \tag{4.1b}$$

where A is the actual cross-section area of the specimen when the force F is applied. The reason for this distinction is that the specimen cross section typically decreases (increases) with applied tensile (compressive) stress. The typical units of stress are megapascal (MPa), which is equal to 10^6 N/m^2, and thousand pounds per square inch (ksi), which is equal to 6.9 MPa.

Upon the application of a tensile (compressive) stress, the specimen becomes longer (shorter). For a specimen of initial length L_o and extension ΔL after initial application of stress, engineering strain ε and true strain ε_T are defined as:

$$\varepsilon = \frac{\Delta L}{L_o}$$

$$\varepsilon_T = \frac{\Delta L}{L}, \tag{4.2}$$

where L ($= L_o + \Delta L$) is the specimen length at a given load. The reason for the distinction between engineering and true strain is that the specimen length changes with applied stress.

At sufficiently small strains, the deformation is elastic; that is, the specimen dimension returns to its original value upon removal of the applied stress. For most materials under these conditions, stress is proportional to strain, as shown in Figure 4.3. The slope of the straight line is known as the elastic or Young's modulus (i.e., $E = \sigma/\varepsilon$ for small ε).

4.2.2 POISSON RATIO

When a material is being stretched along one direction (z-direction), it contracts along the lateral direction (x- or y-direction), as illustrated in Figure 4.4. The ratio of negative lateral strain, $-\varepsilon_L$, to axial strain, ε_A, is the Poisson ratio ν:

$$\nu = -\frac{\varepsilon_L}{\varepsilon_A}. \tag{4.3}$$

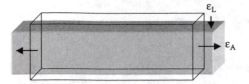

FIGURE 4.4 Illustration of lateral and axial strain.

For typical materials, the Poisson ratio is around 0.25–0.3.

EXAMPLE

For solids in which no volume change occurs after tensile deformation, show that the Poisson ratio is equal to 0.5.

Solution

Consider a specimen block, as shown in Figure 4.4, measuring x, y, and z in three directions. Assume that the applied stress is along the z-direction. The specimen volume $V = x\,y\,z$. Therefore, we can write:

$$\ln V = \ln x + \ln y + \ln z,$$

so that:

$$dV = 0 = \frac{dx}{x} + \frac{dy}{y} + \frac{dz}{z}.$$

Assuming that the x- and y-directions are equivalent, we can write:

$$0 = 2\varepsilon_L + \varepsilon_A;$$

that is, $\varepsilon_A = -2\varepsilon_L$. Therefore, the Poisson ratio is equal to:

$$\nu = -\frac{\varepsilon_L}{\varepsilon_A} = \frac{\varepsilon_L}{2\varepsilon_L} = 0.5.$$

4.2.3 SHEAR STRESS, SHEAR STRAIN, AND SHEAR MODULUS

In Figure 4.2, the two opposing forces applied to the specimen are aligned along the same direction. When they are not aligned, shear deformation occurs (Figure 4.5). In this case, shear stress τ and shear strain γ are defined as follows:

$$\tau = \frac{F}{A};\qquad\qquad\qquad (4.4a)$$

$$\gamma = \tan\theta.\qquad\qquad\qquad (4.4b)$$

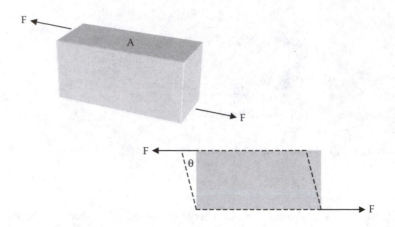

FIGURE 4.5 Specimen test geometry in shear deformation.

Analogous to the definition of Young's modulus, the shear strain is proportional to shear stress for small deformation. The constant of proportionality is the shear modulus G — that is, $G = \tau/\gamma$. It can be shown that, in the linear elastic regime,

$$G = \frac{E}{2(1+v)}.$$
(4.5)

Since the Poisson ratio is on the order of 0.25–0.3, this means that the shear modulus is roughly equal to 0.4 E. For example, the Young's modulus of diamond is about 1000 GPa, while its shear modulus is about 440 GPa.

4.3 BASIC FACTS

4.3.1 YOUNG'S MODULUS

Young's modulus is defined as the initial slope of the stress–strain curve and depends on stress direction. For example, the average value of Young's modulus for Fe is about 205 GPa. However, the value of E is 280 GPa along the (111) direction and 125 GPa along the (100) direction. This is important in considering the deformation of polycrystalline materials. Young's modulus is a measure of the resistance against elastic deformation.

EXAMPLE

Consider a solid made of atoms connected to each other via springs. Assuming Young's modulus of 100 GPa, equilibrium atomic spacing of 0.3 nm, and atom cross-section area of 0.09 nm², estimate the spring constant.

Solution

The force F needed to extend the spring by Δx is equal to $k\Delta x$, where k is the spring constant. We can write:

$$F = k\Delta x = AE\frac{\Delta x}{x_o},$$

where A is the cross section of the atom, E is Young's modulus, and x_o is the equilibrium spacing. The second equality comes from the definition of Young's modulus. Therefore, we have:

$$k = \frac{AE}{x_o}.$$

Note that the spring constant k is proportional to E.* Writing $A = 0.09$ nm^2 and $x_o = 0.3$ nm, we have:

$$k = 0.3 \times 10^{-9} \times 10^{11} = 30 \text{ N/m}.$$

4.3.2 Yield Strength

Yield strength σ_y is a measure of the stress level at which permanent or plastic deformation begins and may be noted as the point of departure from linearity in the stress–strain curve. In most metals and alloys, such departure is gradual. By convention, the yield strength is defined as the stress at the intersection between the stress–strain curve and a straight line parallel to the elastic portion of the stress–strain curve offset at 0.2% strain (Figure 4.6). This is known as 0.2% offset yield strength. Some typical values are:

35 MPa for annealed 1100 aluminum alloys
200 MPa for hot-finished and annealed 304 stainless steels
500 MPa for heat-treated and aged 7075 aluminum alloys
800 MPa for annealed Ti-6Al-4V alloys
1200 MPa for cold-rolled 17-7PH stainless steels

4.3.3 Ultimate Tensile Strength

As denoted in Figure 4.6, the ultimate tensile strength σ_u is the maximum engineering stress that can be sustained by the material. (Reminder: Engineering stress is defined as the force divided by the initial cross-section area of the test specimen.) The stress decrease with further straining beyond this point as shown in Figure 4.6 is somewhat misleading. When the stress approaches the ultimate tensile strength, necking of the specimen occurs, resulting in a smaller cross section. As a result, the true stress (force divided by the actual specimen cross section) is higher than the engineering stress and continues to increase with straining.

---

* Near the minimum of the potential energy curve (similar to Figure 1.8), we can write the energy U of the system in Taylor expansion as:

$$U(x_o + a) = U(x_o) + a \frac{dU}{dx}\bigg|_{x=x_o} + \frac{a^2}{2} \frac{d^2U}{dx^2}\bigg|_{x=x_o},$$

where x_o is the equilibrium position and a the displacement. Since U is at the minimum, the second term on the right-hand side is equal to zero, so:

$$\Delta U = U(x_o + a) - U(x_o) = \frac{a^2}{2} \frac{d^2U}{dx^2}\bigg|_{x=x_o}.$$

Since force $F = d(\Delta U)/dx$, one can deduce that $F = ka$, where $k = d^2U/dx^2$. As shown in this example, k is directly related to the Young's modulus. Therefore, in solids without voids or massive defects, the Young's modulus E is a measure of the curvature of the potential energy curve at the equilibrium spacing between atoms.

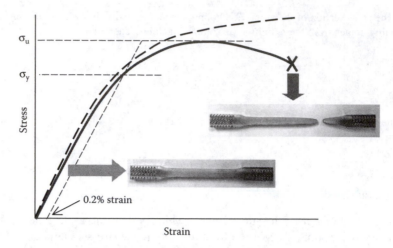

FIGURE 4.6 Definition of 0.2% offset yield strength (σ_y) and ultimate tensile strength (σ_u) in an engineering stress versus engineering strain plot. Note the necking of failed specimen. The dotted line represents the same plot for true stress versus true strain. (Micrographs reprinted with permission from Kathleen Stair.)

4.3.4 PLASTIC STRAIN

From the stress–strain curve, we can obtain the total strain ε_{tot} at a given applied stress σ. The total strain consists of elastic strain, ε_{el}, and plastic strain, ε_{pl}, given by:

$$\varepsilon_{tot} = \varepsilon_{el} + \varepsilon_{pl} = \frac{\sigma}{E} + \varepsilon_{pl}. \tag{4.6}$$

Figure 4.7 shows how elastic strain is separated from plastic strain at a given stress.

4.3.5 HARDNESS

Hardness is a measure of resistance against plastic deformation. Many applications require materials to be wear resistant. Under some conditions, wear resistance is directly related to hardness.* Hardness is measured by several indentation techniques. The two most popular methods are Rockwell and Vickers indentation measurements for macroscopic specimens. The Rockwell technique involves indenting a sample with a diamond cone. Within this technique, there are different Rockwell scales. The most common is Rockwell C. In this case, the difference in indentation depths induced by 150- and 10-kg loads is measured. The smaller the difference is, the larger the Rockwell C hardness number will be.

The other method is the Vickers microhardness test. In this case, the sample is indented with a four-sided diamond pyramid at a given load P and then the total area in contact, A_{total}, is measured with the diamond indenter. Experimentally, we measure the area projected in the direction of applied load A_{proj} — that is, $A_{proj} = A_{total} \sin 68°$ (68° is one half the full opening angle of the pyramidal indenter). The Vickers hardness H_V value is defined by:

$$H_V = \frac{P}{A_{total}} = \frac{0.927P}{A_{proj}}. \tag{4.7}$$

---

* The science of friction, wear, and lubrication is known as tribology (*tribos* is a Greek word meaning rubbing). In abrasive contacts, there is a strong correlation between wear resistance and hardness. In tribological contact situations in which wear is controlled by lubrication and surface chemistry, such correlation does not exist.

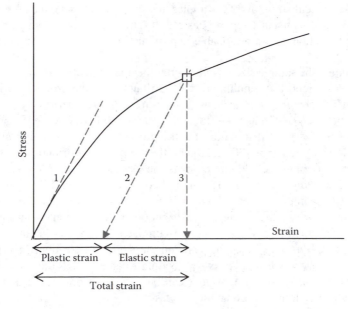

FIGURE 4.7 Determination of elastic and plastic strain. Step 1: Draw the initial slope of the stress–strain curve. Step 2: Starting from the point of interest (□), draw a line parallel to the first line in Step 1 to intercept the strain (*x*) axis. Step 3: From the point of interest, draw a line parallel to the stress (*y*) axis to intercept the strain (*x*) axis. The total strain, elastic strain, and plastic strain are shown.

The 0.927 factor comes from sin 68°, and the unit is kilograms per square millimeter (9.8 MPa).*

To study mechanical properties of thin films, a technique similar to the Vickers indentation method is used, with three differences. First, much lighter loads (typically in the millinewton range or smaller, compared with newton range in standard Vickers measurements) are used to minimize effects from the substrate. Second, the indenter is a three-sided diamond pyramid (known as the Berkovich indenter). The reason for using a three-sided pyramid is that it is easier to polish a tip so that three surfaces converge to a point. In this way, a sharper and better defined tip can be obtained. Third, because the size of an indent is small when studying thin films, the tip penetration depth is used to deduce the indented area. Therefore, the instrumentation must be sensitive to displacements in the nanometer range. Hence, this technique is known as nanoindentation. The nanoindentation hardness H_{nano} is defined without the sine factor as follows:

$$H_{nano} = \frac{P}{A_{proj}}.$$ (4.8)

Statistics Notes

Because of the sensitivity to small compositional and microstructural changes, mechanical properties of materials are subject to some degree of statistical spread due to slight changes in processing conditions. The degree of statistical spread should always be reported. The standard approach taken by most engineers and scientists is to measure the property of interest

---

* The hardest substance in nature is diamond. The hardness of diamond is about 90–110 GPa. Cubic boron nitride is the next hardest, with hardness in the range of 50–70 GPa. In comparison, the hardness of carburized or case-hardened steels is around 10 GPa.

(such as hardness or yield strength) of n samples and then compute the mean, μ, and standard deviation, σ. The result is usually reported as $\mu \pm \sigma$. This reporting scheme assumes that the sample standard deviation σ is a reasonable approximation of the standard deviation for the entire lot.

Clearly, the larger the sample size n is, the better the approximation is. It is therefore useful to include n in such a report. Assuming normal distribution for the property of interest in the lot, there is a 68% probability that any *single* measurement falls between $\mu - \sigma$ and $\mu + \sigma$, 95% probability between $\mu - 2\sigma$ and $\mu + 2\sigma$, and 99% probability between $\mu - 3\sigma$ and $\mu + 3\sigma$.

The assumption of normal distribution for the property of interest is not always correct. When the sample size n is large (30 or larger), the frequency distribution of the computed means μ for these n samples should approximate the Gaussian or normal distribution having standard deviation σ_μ equal to $\sigma/\sqrt{n-1}$. This is known as the central limit theorem. In this situation, 95% of the computed mean should fall between $\mu - 1.96\,\sigma_\mu$ and $\mu + 1.96\,\sigma_\mu$.

When n is small (<30), the frequency distribution of the computed means follows the t-distribution instead. In this case, 95% of the computed mean should fall between $\mu - t\,\sigma/\sqrt{n-1}$ and $\mu + t\,\sigma/\sqrt{n-1}$, where t depends on the degree of freedom ($n - 1$). When $n = 10$, the t-value is 2.25; when $n = 5$, the corresponding t-value is 2.78 (contrast this with 1.96 for normal distribution). Appropriate t-values can be obtained from standard statistics tables for other degrees of freedom and probability values.

Public opinion polling uses similar strategies. In this way, it becomes possible to estimate the collective opinions of millions of people by polling only several hundred randomly chosen people, with reasonably small statistical uncertainty or percentage margin of error. For example, when N individuals are asked if they like a certain candidate, a certain fraction p says yes. The standard deviation is $\sqrt{Np(1-p)}$, and the percentage margin of error for one standard deviation is $\sqrt{(p(1-p)/N}$. Therefore, when 500 randomly chosen individuals are polled and 60% say that they will vote for a certain candidate, the percentage margin of error for one standard deviation is then about 2%. This means that there is 95% probability that this candidate will win between 56 and 64% of the vote. Of course, opinions are time dependent; polling results are strictly valid on the day of the polling and only if the sampling is truly random.

EXAMPLE

We have made a large batch of a given material. Its average hardness is 5.0 GPa, and the standard deviation is 0.6 GPa. If 10 random samples are taken from this batch, determine the average hardness with a confidence interval of 95%.

Solution

When $n = 10$, the t-value is 2.25 for 95% confidence interval. Therefore, with 95% probability, the average hardness should be equal to:

$$5.0 \pm \left(2.25 \times \frac{0.6}{\sqrt{10-1}} \right) \Rightarrow 5.0 \pm 0.45 \text{ GPa.}$$

4.4 PLASTIC DEFORMATION

When a metal is stressed beyond its elastic limit, we often see the formation of slip bands. Such slip bands are formed by the sliding of one plane of atoms over another (Figure 4.8a). There are two key facts:

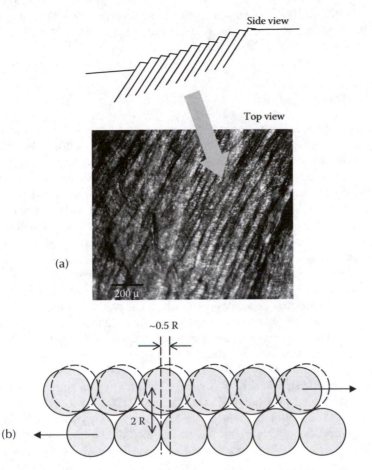

FIGURE 4.8 (a) Formation of slip bands during tensile deformation, illustrated by a schematic side view and a top view of a zinc sample. Each slip band is formed by the sliding of atoms along a specific plane. (Micrograph reprinted with permission from Kathleen Stair.) (b) Sliding of one plane over another. The critical displacement to obtain plastic or permanent deformation is about $0.5R$, where R is the atomic radius. Since the distance between the two planes is about $2R$, the critical shear strain for plastic deformation is therefore equal to $0.5R/2R$, or about 1/4.

1. Crystals have well-defined slip planes and slip directions. A slip system specifies the slip plane (hkl) and the slip direction $[uvw]$. For example, for face-centered cubic crystals, the slip planes are $\{111\}$ and the slip directions are $<110>$. In general, the most favorable slip plane is the closest packed plane. The most favorable slip direction is one with the highest linear density of atoms. We will defer the explanation for these preferred slip systems to the next section.*

2. When we slide one plane of atoms over another, the displacement needed to slide the upper plane of atoms (relative to the lower plane) up to the steepest part of the slope is about 0.5 the atomic radius (Figure 4.8b). Once that is accomplished, plastic or permanent deformation occurs. This corresponds to a shear strain of ~0.5/2 = 1/4. Therefore, for a perfect crystalline solid with shear modulus G, the minimum or critical shear stress τ_c required to cause slip is $\approx G/4$. Experimentally, τ_c for most engineering alloys is between

---

* FCC crystals have more slip systems available than BCC or HCP crystals. As a result, materials with the FCC structure are usually more ductile than their BCC or HCP counterparts. This is an important consideration in alloy design for specific applications.

$G/100$ and $G/1{,}000$. The explanation for this large discrepancy is that, in real solids, we do not have perfect atomic planes sliding over each other. Rather, the presence of dislocations decreases the critical stress needed for plastic deformation, as elaborated further in the next section.

4.5 DISLOCATIONS

The simple analysis presented in the preceding paragraph indicates that, in a perfect crystalline solid, it is difficult to slide one plane of atoms over another. We can trace the reason to the fact that we have to move all atoms on the slip plane at the same time. In real solids, there are line defects known as dislocations as discussed in Chapter 2. Recall our discussion, for example, on edge dislocations. When shear stress is applied, only one column of atoms has to be moved at any given time, instead of all atoms on the slip plane. Therefore, the critical shear stress to cause permanent deformation τ_c is much smaller than expected of dislocation-free solids.

It has been shown that the energy associated with a line dislocation (edge or screw) is proportional to Gb^2 per unit length of dislocation, where G is the shear modulus and b the length of the Burgers vector. Therefore, dislocations responsible for plastic deformation are those with the smallest shear modulus and shortest Burgers vector (to obtain minimum dislocation energy — based on one of the fundamental principles discussed in Chapter 1). In order to have the smallest shear modulus, the shear planes should be widely separated and must therefore be the closest packed.* To have the shortest slip vector, the slip direction must contain the highest atomic density. These arguments are consistent with experimental findings.

It is important to note that oddly sized substitution atoms decrease stress around dislocations and hence lower the elastic strain energy. As discussed earlier, given sufficient mobility, impurity atoms tend to collect around dislocations. These regions are known as Cottrell atmospheres and are energetically stable. Therefore, it is necessary to supply energy to detach dislocations from these solute or impurity atoms. This means that the addition of solute or impurity atoms results in strengthening. This phenomenon is known as *solid solution strengthening* (Figure 4.9) and provides an opportunity to control material strength via introduction of impurity atoms. In the example shown in Figure 4.9, copper–nickel alloys are stronger than either of the constituent elements.

Plastic deformation increases the number of dislocations. These dislocations may repel and interfere with one another. As a result, materials become stronger after deformation.** This phenomenon is known as *work or strain hardening* (Figure 4.10). After work hardening, the ductility

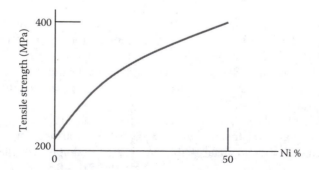

FIGURE 4.9 Tensile strength of copper–nickel alloys versus Ni concentration.

---

* Since the number of atoms per unit volume is a constant within the material, a stack of closely packed planes implies that the spacing between these planes is large.
** A dislocation-free solid is strong. Having lots of dislocations also makes a solid strong. Therefore, this suggests that a solid is weakest at some intermediate dislocation density.

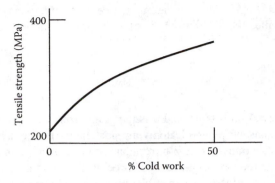

FIGURE 4.10 Tensile strength of copper as a function of percent cold work.

is reduced and the material is prone to brittle fracture (see discussion in a subsequent section on toughness). When numerous dislocations are present and the material is subjected to high stress, dislocations of the same sign will eventually pile against some obstacle (e.g., hard precipitates or grain boundaries) and form an internal void. Voids may act as initiation sites for failure due to stress concentration.

When dislocations encounter soft precipitates, they pass through them. Hard precipitates, on the other hand, act as obstacles against dislocation motion. Theory shows that dislocations loop around these obstacles. The shear stress needed to form dislocation loops around hard precipitates is inversely proportional to the distance between precipitates. Therefore, the yield strength increases with precipitate concentration. This phenomenon is known as *precipitation hardening*.

4.6 PLASTIC DEFORMATION OF POLYCRYSTALLINE MATERIALS

Most engineering alloys are polycrystalline — that is, they are made up of single-crystal grains at different orientations separated by boundaries. Because of this variation in orientation, each grain is strained differently in the presence of applied stress. For example, consider the deformation of a single crystal grain with geometry shown in Figure 4.11. The stress axis is at an angle θ from the surface of the slip plane. The slip direction is at an angle λ from the plane containing the stress axis and the normal of the slip plane. The shear stress resolved along the slip direction σ_R is given by:

$$\sigma_R = \sigma \cos \theta \cos \lambda \,, \tag{4.9}$$

where σ is the applied stress. The quantity σ_R is known as resolved shear stress, which is therefore different for grains of different orientations.

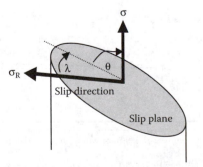

FIGURE 4.11 Deformation of a single crystal.

At low temperatures (<0.3–0.5 of the melting temperature in kelvins), small-grained materials are stronger. For example, the yield strength σ_y of a polycrystalline material is given by the Hall–Petch relationship:

$$\sigma_y = \sigma_o + kd^{-1/2}, \tag{4.10}$$

where σ_o and k are materials constants, and d is the grain size.

There are three reasons why materials having smaller grain size are stronger. First, since a grain boundary represents a transition region from one single-crystal grain to another with different orientation, the slip direction that is favorable in one grain will not be favorable in an adjacent grain. As such, grain boundaries act as obstacles to dislocation motion. Second, because of the variation of θ and λ in different grains, some grains are less strained, and there is less dislocation propagation through these grains. Third, in the limit when the grain size is less than 10 nm, molecular dynamics simulation studies show that it is energetically unfavorable to have dislocations in these grains. As discussed earlier, dislocation-free solids are very strong.

4.6.1 CREEP

If you hang a weight onto a rubber band and note the position of the weight over a period of several hours, you will notice that the rubber band becomes longer and longer. In other words, even when the load or stress is constant, strain increases with time. This phenomenon is known as creep and is shown in Figure 4.12.

Creep is due to atomic diffusion and is active at temperatures greater than or equal to 0.3–0.5 of the melting point in kelvins of the material. At these temperatures, atoms readily diffuse along grain boundaries. Figure 4.13 shows that when stress is applied as shown, atoms at grain boundaries oriented more or less parallel to the stress axis are under compression, while atoms at grain boundaries oriented perpendicular to the stress axis are under tension. To reduce the elastic strain energy associated with compression and tension in various grain boundaries, atoms migrate from grain boundaries under compression (which are approximately parallel to the stress axis) to grain boundaries under tension (which are oriented approximately perpendicular to the stress axis). This results in elongation of the sample and is known as Coble creep.

In addition to atomic diffusion along grain boundaries, another mechanism is responsible for creep at high temperature, due to the interaction between vacancies and dislocations. This is known as dislocation climb. Consider Figure 4.14 showing a vacancy in the vicinity of an edge dislocation, as denoted by the inverted "T." Sequential atomic diffusion causes the glide plane of the dislocation to climb as shown. With sufficient vacancy concentration, the dislocation can climb far away from the original glide plane. This reduces stress on the original glide plane, which can then accommodate injection of more dislocations at constant stress, thus resulting in creep.

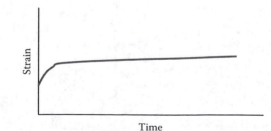

FIGURE 4.12 The phenomenon of creep.

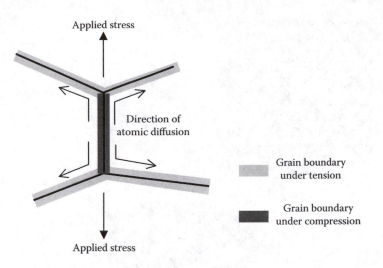

Applied stress

Direction of
atomic diffusion

Grain boundary
under tension

Grain boundary
under compression

Applied stress

FIGURE 4.13 Atomic diffusion along grain boundaries under stress.

In certain high-temperature applications, such as turbine blades used in jet engines, this change in sample dimension is not acceptable. Bulk diffusion is known to be much slower than grain boundary diffusion. Therefore, one way to minimize creep is to eliminate grain boundaries with the use of single-crystal alloys. That is why turbine blades in jet engines are made of nickel aluminide single-crystal alloys.

The creep strain rate $\dot{\varepsilon}$ can be written as:

$$\dot{\varepsilon} = K\sigma^n \exp\left(-\frac{Q}{k_B T}\right),$$ (4.11)

where

σ = the applied stress
Q = the activation energy for creep
T = absolute temperature
k_B = Boltzmann constant
K and n = constants

Therefore, the time to failure or rupture time, t_r, should be proportional to the inverse of the right-hand side of Equation (4.11):

$$t_r = \frac{1}{K'\sigma^n} \exp\left(\frac{Q}{k_B T}\right).$$ (4.12)

Hence, it can be shown that:

$$\frac{Q}{2.3k_B} = T(\log_{10} t_r + \log_{10} K' + n\log_{10}\sigma) = T(\log_{10} t_r + C).$$ (4.13)

At a given stress, the parameter $T(\log_{10} t_r + C)$ is a constant for a given system and is known as the Larson–Miller parameter. Once this parameter is known at a given stress, we can determine the rupture time as a function of temperature at the same stress.

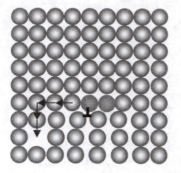

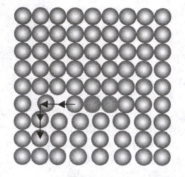

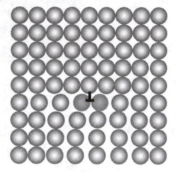

FIGURE 4.14 Illustration of dislocation climb.

EXAMPLE

The Larson–Miller parameter for a given alloy is 2.5×10^4. At 1000 K, the rupture time is 2000 h at a given stress. How long can this alloy survive at 1100 K at the same stress?

Solution

Equation (4.13) can be written as:

$$T_1(\log_{10} t_1 + C) = T_2(\log_{10} t_2 + C).$$

Therefore,

$$1000(\log_{10} 2000 + C) = 2.5 \times 10^4 = 1100(\log_{10} t_2 + C).$$

From the first equality, the quantity C is calculated to be 21.7. Substituting this into the second equality gives 10.6 h for t_2. Note that a small temperature increase results in a large decrease in rupture time.

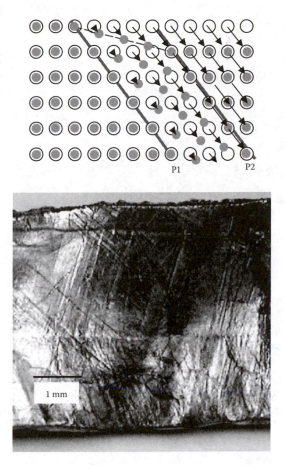

FIGURE 4.15 Deformation by twinning. The open circles represent original atomic positions, while the shaded circles represent atomic positions after deformation. Note the mirror symmetry of positions of displaced atoms relative to the other atoms across the twinning planes as denoted by P1 and P2. The accompanying micrograph shows a tin sample after deformation. Note the cross-hatched bands near the middle of the picture. (Micrograph reprinted with permission from Kathleen Stair.)

4.6.2 CRYING TIN

Compared with FCC metals, metals with the BCC or HCP structure have fewer slip systems available for plastic deformation. Under appropriate deformation conditions, another deformation mechanism known as twinning becomes operative before the activation of dislocation motion. This twinning mechanism operates by the collective motion of atoms, resulting in their being located in mirror image positions of other atoms, as shown in Figure 4.15. Note the mirror symmetry of positions of displaced atoms relative to the other atoms across the twinning planes as denoted by P1 and P2. When twinning occurs during plastic deformation of HCP tin, it is accompanied by characteristic acoustic emission — hence, the term "crying tin."

4.7   RECOVERY OF PLASTICALLY DEFORMED METALS

After cold working, the material contains many defects and dislocations. This results in substantial stored strain energy. The system can eliminate this excess energy by atomic diffusion, which is sufficiently rapid at temperatures about 0.3–0.5 of the melting point (in kelvins). This heat treatment or annealing process results in defect removal and recovery of mechanical properties, such as yield

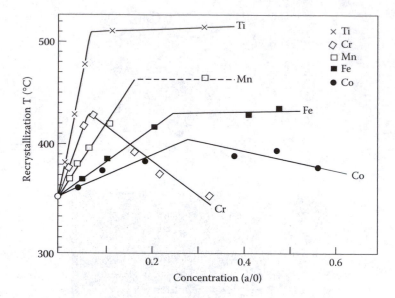

FIGURE 4.16 Effect of selected impurities on the recrystallization temperature of Ni. (Adapted from E.P. Abramson II. Dilute alloying effects on recrystallization in nickel as compared with other transition element solvents, *Transactions of the Metallurgical Society of AIME* 224:727, 1962.)

strength, toughness, and ductility. Since the driving force for recovery is the stored strain energy, the recovery rate depends on the amount of plastic deformation. In addition, since the rate of atomic diffusion depends on material composition, the rate of recovery depends on temperature and sample purity. Figure 4.16 shows the effect of impurities on the recovery of Ni, as measured by the recrystallization temperature. Note that it does not take a lot of impurities to cause a large change in the rescrystallization temperature.

4.8 FRACTURE

4.8.1 TOUGHNESS

As shown in Figure 4.17, there are two types of fracture failure: brittle and ductile. The shaded area under the stress–strain curve represents the energy required to fracture the material and is a

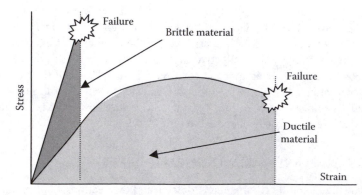

FIGURE 4.17 Stress–strain curves for brittle and ductile materials. Note the absence of yielding or plastic deformation for brittle materials. Also, the area under the stress–strain curve up to failure, which is a measure of the energy per unit volume required for failure, is smaller for brittle materials.

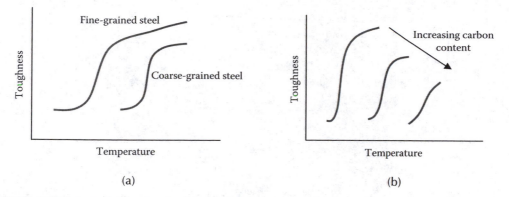

FIGURE 4.18 Schematic diagram showing (a) the dependence of toughness of steels on grain size; and (b) how carbon concentration in steel affects toughness as a function of temperature.

measure of toughness.* Generally, all materials have lower toughness at lower temperature. FCC metals have multiple slip systems, and the atoms are closely packed so that slip occurs easily with minimal thermal activation. Consequently, FCC metals tend to be more ductile, and their toughness is less sensitive towards temperature.** Metals with BCC, HCP, and other structures have fewer slip systems that can be activated at room temperature. For these metals, a ductile-to-brittle transition temperature exists below which toughness decreases rapidly.

Furthermore, toughness depends on grain size and impurity concentration.*** Figure 4.18(a) indicates that smaller grain size results in higher toughness. According to the Hall–Petch relationship, smaller grain size also results in higher strength. This is one of the few instances in which strength and toughness can be increased at the same time. Figure 4.18(b) shows the effect of carbon on the toughness of steels: Increased carbon concentration decreases the toughness and raises the ductile-to-brittle transition temperature. A tough material deforms plastically with cracks propagating in a stable manner. It absorbs large amounts of strain energy and gives ample warning before failure. A brittle material such as glass or ceramic, on the other hand, sustains high stress with little or no plastic deformation (due to limited dislocation motion) and fails without warning when the ultimate strength is exceeded.

4.8.2 Fracture Mechanics

The theoretical cohesive strength of a material should be on the order of $E/10$, where E is the elastic modulus. Experimental values are usually in the range of $E/100$ to $E/10,000$. As discussed in an earlier chapter, the presence of dislocations in real solids is one explanation for the lower than expected strength. Another explanation proposed by A.A. Griffith is the presence of flaws (surface and bulk). These flaws amplify the nominal stress to higher values. For example, the stress σ in the vicinity of an elliptical crack (Figure 4.19) of length $2a$ and local radius r is given by:

$$\sigma = \sigma_o \left[1 + 2 \left(\frac{a}{r} \right)^{1/2} \right],$$
(4.14)

---

* One popular method to measure toughness is the Charpy impact test. In this test, we start with a specimen notched and mounted in a standard configuration. A pendulum carrying a "hammer" at a certain initial height impacts the specimen, fractures it, and ends at a final height. The height difference gives the energy used to fracture the sample and provides a relative measure of material toughness.

** Iridium has the FCC structure and is quite brittle at room temperature. The exact reason is not known.

*** Coarse-grained steels lose their toughness at low temperatures faster than their fine-grained counterparts. It has been suggested that hulls of early oil tankers were made of coarse-grained steels. The loss of quite a few of these tankers in the North Sea may be due to hull fracture in the cold and stormy ocean.

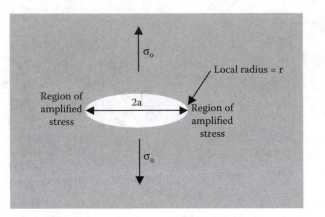

FIGURE 4.19 Regions of amplified stress near a crack or flaw.

where σ_o is the externally applied stress. The factor,

$$1 + 2 \left(\frac{a}{r} \right)^{1/2},$$

can be considered to be the stress amplification factor by these flaws. Therefore, the revised failure criterion is:

$$\sigma_o \left[1 + 2 \left(\frac{a}{r} \right)^{1/2} \right] \geq \frac{E}{10}. \tag{4.15}$$

This equation applies to materials that do not exhibit plasticity (e.g., ceramic materials), so the local radius of curvature r is unchanged in the presence of external stress. For materials that exhibit plasticity at high stresses, plastic deformation causes blunting of the initial crack (i.e., r is not a constant).

For a preexisting crack to propagate, we need a certain critical stress. Griffith analyzed this problem by balancing the energy needed to create new surfaces with the release of elastic strain energy during crack propagation. For a thin brittle solid, he arrived at the following expression for the critical stress σ_c required for crack propagation:

$$\sigma_c = \left(\frac{2 E \gamma_s}{\pi a} \right)^{1/2}, \tag{4.16a}$$

where γ_s is the surface energy per unit area (i.e., energy required to create new surfaces).

For materials that exhibit plastic deformation before fracture, the preceding equation is modified by replacing the term γ_s with $\gamma_s + \gamma_p$, where γ_p is the energy associated with plastic deformation in extending the crack. For ductile materials, γ_p is the dominant term. In this case, writing G as $2(\gamma_s + \gamma_p)$, we have:

$$\sigma_c = \left(\frac{GE}{\pi a} \right)^{1/2}. \tag{4.16b}$$

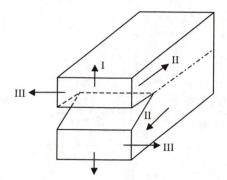

FIGURE 4.20 Three crack-opening modes.

TABLE 4.1
Mode I Fracture Toughness for Selected Alloys, Ceramics, and Polymers

Materials	K_{Ic} (MPa-m$^{1/2}$)
7075-T651	24
2024-T3	44
Ti-6Al-4V	55
4340 tempered @425°C	87
Concrete	0.2–1.4
Soda-lime glass	0.7–0.8
Alumina	2.7–5.0
Polycarbonate	2.2

The preceding analysis shows that the parameter controlling crack propagation is proportional to $\sigma(\pi a)^{1/2}$. For a given test geometry, we define the quantity K as the stress intensity factor:

$$K = Y\sigma\sqrt{\pi a} \, , \tag{4.17}$$

where Y (≈ 1) depends on the crack geometry and the testing configuration.

Note that for surface cracks, a is the crack depth. The critical value for crack propagation K_c is known as the fracture toughness.* The three testing configurations resulting in different crack-opening modes are shown in Figure 4.20.

The three crack-opening modes are plane strain (mode I), plane shear (mode II), and antiplane shear (mode III). Associated with each mode is a different fracture toughness value. Fracture toughness values (K_{Ic}) for mode I cracks for several materials are given in Table 4.1. Note the relatively low fracture toughness of engineering ceramics and plastics compared with metals.

EXAMPLE

Consider a material with K_c = 50 MPa-m$^{1/2}$. Assume that the geometric factor Y is equal to 1. In the presence of a 5-mm deep surface crack, calculate the critical stress σ_c for failure.

---

* In plane-stress condition (i.e., the stress occurs only on a plane), $G = K^2/E$. In plane-strain condition, $G = K^2 (1 - \nu^2)/E$.

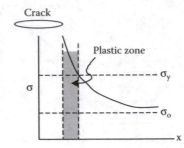

FIGURE 4.21 Stress distribution in front of a crack.

Solution

For $Y = 1$, $\sigma_c = K_c / \sqrt{\pi a}$. In this case, $a = 5$ mm. Therefore, the critical stress σ_c is given by:

$$\frac{50}{\sqrt{0.005\pi}} = 399 \text{ MPa.}$$

When the applied stress is greater than σ_c, the crack propagates to failure.*

Generally, an inverse relationship is found between strength and toughness; stronger materials tend to have lower toughness. This inverse relationship arises from the size of the plastic zone ahead of a crack tip. Consider an elliptical crack of length $2a$ with an external stress σ_o applied as shown in Figure 4.21. The resulting stress shows a mathematical singularity at the crack tip and decreases with distance from the crack tip as follows:

$$\sigma(x) = \frac{\sigma_o}{\sqrt{1 - \dfrac{a^2}{x^2}}}. \tag{4.18}$$

As can be seen from Figure 4.21, the higher the yield strength, the smaller the plastic zone. A smaller plastic zone means that less energy is required to propagate through this zone. Since toughness is related to the amount of energy needed to propagate the crack (Figure 4.17), this indicates that, if other things are equal, a high-strength material generally has lower toughness.

Question for discussion: Thus far, we have presented the following methods to improve material strength: solid solution strengthening, work hardening, small grain size, and precipitates. Discuss how each method affects toughness.

4.8.3 FATIGUE

When a material is subjected to cyclic loading, failure can occur even when the applied stress is well below the ultimate tensile strength of the material. This mode of failure is known as fatigue and is the most common mode of failure in dynamical structures (aircraft, machine components, bridges, etc.). One approach to study this phenomenon is to apply cyclic loading with a given stress range to a material in a given geometry (as shown in Figure 4.22) and measure the number of

---

* Failure due to small flaws is a major concern, especially in materials with low fracture toughness. To detect these subsurface defects requires the use of nondestructive testing techniques such as x-ray or ultrasonic imaging.

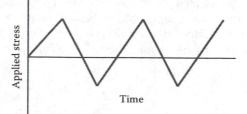

FIGURE 4.22 An example of cyclic loading.

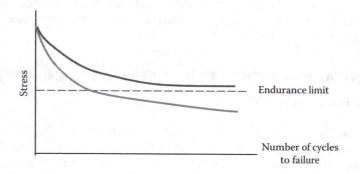

FIGURE 4.23 Examples of S–N curves: dependence of fatigue life (number of cycles to failure) on the applied maximum stress with and without an endurance limit.

cycles for failure to occur. The typical test is such that the minimum stress is zero or equal to minus maximum stress (fully reversed fatigue test). These measurements are repeated for different maximum stress levels. The resulting data, known as the S–N curve, are plotted in Figure 4.23. In some cases, an asymptotic limit is obtained below which infinite life is attained, as illustrated in Figure 4.23. This is known as the endurance limit.

In some cases, the endurance limit does not exist. The S–N curve allows materials engineers to design structures with known life or fail-safe criterion. Note that stress can also arise from thermal expansion mismatch. For example, in circuit boards of large computer systems, silicon chips are soldered directly onto ceramic substrates. The solder material has a larger thermal expansion coefficient than silicon or the ceramic. When the system undergoes heating and cooling cycles, the solder joint is subjected to thermomechanical fatigue, which may lead to the formation of fatigue cracks and eventual failure.

The basic mechanism of fatigue is slip irreversibility. When the material is subject to shear, slip occurs, resulting in a slip step of certain length. In the reverse cycle, the slip is not fully reversed along the original slip step. Part of the reverse slip occurs at a different location, resulting in the formation of a permanent depression, as shown in Figure 4.24. Research studies indicate that the

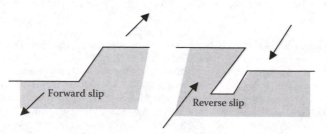

FIGURE 4.24 Generation of a small depression due to slip irreversibility, leading eventually to the formation of a fatigue crack.

degree of slip reversibility depends on material and environment. Repetition of forward and reverse slip results in the initiation of a fatigue crack, which then propagates with further fatigue cycling. Without distinguishing between initiation and propagation, the fatigue life to failure N_f is related to the maximum plastic strain ε_p by the Coffin–Manson relationship:

$$N_f = \alpha \varepsilon_p{}^\beta, \qquad\qquad (4.19)$$

where α and β are materials parameters. β is typically between -0.5 and -1.0.

Many "unexpected" fatigue failures occur because of poor quality control (such as large inclusions in the material), poor surface finish, and unintentional stress concentrators in component design. Since fatigue crack initiation and propagation involve the generation of new surfaces, fatigue life depends on the environment.*

4.9 MECHANICAL PROPERTIES, SURFACE CHEMISTRY, AND BIOLOGY

4.9.1 Fatigue Life of Metals

As noted earlier, the fatigue life of metals and alloys depends on the environment. When steel is subjected to fatigue cycling, slip steps consisting of fresh metal surfaces are exposed. In air, ambient gas molecules (e.g., oxygen) adsorb on these slip steps and are bound strongly. The presence of strongly bound adsorbates decreases the slip reversibility, accelerating the buildup of fatigue damage. At moderate strain amplitudes ($<10^{-3}$), the fatigue life of steels is reduced by as much as a factor of 10 in air compared with vacuum.

4.9.2 Ductility of Nickel Aluminide

Another example is the tensile ductility of polycrystalline Ni_3Al, which is a common material used in turbine blades. At moderate strain rates ($<10^{-3}$/s), the fracture strain of polycrystalline Ni_3Al is about 0.5% in ambient air. However, in dry oxygen, the fracture strain is about 10–20%. In ultrahigh vacuum (pressure less than 1×10^{-9} torr), its fracture strain increases to about 50%. The culprit is water vapor. During tensile testing, small cracks open up along grain boundaries, exposing fresh metal surfaces. Water vapor adsorbs and interacts with these surfaces, producing atomic hydrogen (Figure 4.25). Atomic hydrogen then diffuses rapidly along the grain boundaries to the crack tip, causing embrittlement of the material. Removal of water vapor eliminates the embrittlement problem.

4.9.3 Tin Whiskers

The growth of tin whiskers is an annoying problem in electronics. Tin is used as a solder joint material. Over time, whiskers are observed to grow from the tin surface; they lead to short circuiting of the electronics, thus creating reliability problems. This is especially serious for electronics used in unmanned space vehicles such as satellites, in which repair cannot be done easily. For a long

---

* Fatigue failures account for many aircraft accidents. In the early days of aviation, square windows were used that resulted in a stress concentration around the corners and subsequent fatigue cracking. Fatigue was also the cause of failure in Aloha Airlines flight 243 on April 28, 1988, according to the final report of the National Transportation Safety Board. The aircraft experienced structural failure and explosive decompression at 24,000 ft while en route from Hilo to Honolulu. Approximately 18 ft of cabin skin and structure aft of the cabin entrance door and above the passenger floorline separated from the aircraft. One flight attendant who was standing in the aisle was swept overboard. The flight diverted to Maui and landed without further incident. Subsequent examinations of the aircraft revealed fatigue damage of certain joints. Each takeoff–landing cycle subjected the fuselage to cyclic stresses. Since this plane flew frequent commuter flights among the islands of Hawaii, it was subjected to an unusually large number of fatigue cycles. In addition, the saltwater environment probably accelerated fatigue crack propagation and hence failure.

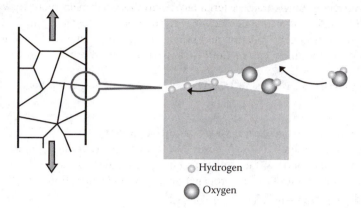

FIGURE 4.25 Illustration of moisture-induced embrittlement of polycrystalline Ni$_3$Al.

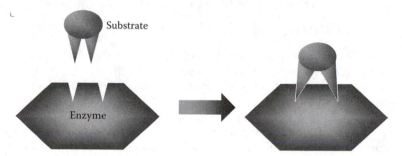

FIGURE 4.26 Induced-fit model for enzyme–substrate interaction.

time, researchers speculated that this was due to some sort of creep phenomenon, given that tin has a low melting temperature. Then it was discovered that oxygen is the problem. Oxidation of tin creates compressive stress within the material, causing tin atoms to migrate outward and resulting in the formation of tin whiskers. We might call this oxidation-induced creep. Whisker growth can therefore be suppressed or eliminated by the removal of oxygen from the environment.

4.9.4 ENZYMES

Enzymes are proteins that accelerate chemical reactions essential for life. A protein molecule consists of one or more chains of amino acids. The chain folds upon itself to generate a particular shape or conformation. This shape allows a given reactant molecule (known as the substrate in biochemistry) to bind to a specific active site of the enzyme — hence, the specificity of enzyme-catalyzed reactions. Active sites are grooves on the enzyme surface.

There are two models describing the substrate–enzyme interaction. One is the lock-and-key model, in which the substrate fits exactly into the active site. The second model is the induced-fit model, in which the substrate deforms slightly to fit into the active site (Figure 4.26). Such distortion may provide a better geometric configuration for the subsequent chemical reaction, or it may facilitate the breaking of certain chemical bonds. The net effect is that such mechanical distortion lowers the activation energy, thus accelerating the overall chemical reaction.

EXAMPLE

The amount of distortion does not have to be large to have dramatic effects. Using the example in Section 4.3, we may assume the equivalent spring constant k in typical chemical bonds to be 30 N/m. With bond length distortion Δx of 0.02 nm (about 10% of interatomic spacing) for 10 bonds, calculate the total energy change. If this translates into reduction of the activation energy, calculate the rate increase at 37°C.

Solution

The total energy change ΔE is equal to $1/2k(\Delta x)^2 N$, where N is the number of bonds affected. This is given by:

$$\frac{1}{2} 30 \times \left(2 \times 10^{-11}\right)^2 \times 10 = 6 \times 10^{-20} J$$

$$\approx 0.375 \text{ eV}.$$

At 37°C (310 K), the Boltzmann factor increases by:

$$\exp\left(\frac{\Delta E}{k_B T}\right) = \exp\left(\frac{6 \times 10^{-20}}{1.38 \times 10^{-23} \times 310}\right)$$

$$\approx 1.23 \times 10^6,$$

where k_B is the Boltzmann constant. The small distortion results in a million-fold acceleration of the chemical reaction.

The message is clear: There is strong interplay between chemistry and mechanical properties. In a chemically active environment, it is no longer sufficient to think about materials by themselves; how they may interact with their surrounding environments must also be considered. This is especially significant with micro- or nano-sized materials, in which the larger surface-to-volume ratio means surface or interface chemistry may contribute significantly to overall mechanical properties.

4.10 MATERIALS SELECTION: MECHANICAL CONSIDERATIONS

Materials are chosen for a specific structural application for various reasons: mechanical properties, cost, availability, health/environmental concerns, etc. In this section, we will provide one example in which the choice is made according to mechanical properties.

Consider the choice of materials for a wire of length L and radius r (Figure 4.27) that is used to support a load W. The requirements are that (1) the strain must not exceed a certain maximum value ε_{max}; and (2) the mass of the wire must be as small as possible. Therefore, we can write:

$$\frac{W}{\pi r^2} < E\varepsilon_{max}, \tag{4.20}$$

where E is the elastic modulus of the wire material. The mass of the beam m is:

$$m = \rho(\pi r^2 L), \tag{4.21}$$

where ρ is the mass density of the beam material. From Equation (4.20) and Equation (4.21), we obtain:

$$m > \rho L \frac{W}{E\varepsilon_{max}} = \frac{LW}{\varepsilon_{max}} \frac{\rho}{E}. \tag{4.22}$$

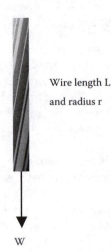

Wire length L

and radius r

W

FIGURE 4.27 Wire under load.

TABLE 4.2
Elastic Modulus and Density for Selected Elements

Metal	Elastic Modulus E (GPa)	Density ρ (g/cm³)
Al	71	2.7
Cr	238	7.1
Co	203	8.7
Cu	123	8.9
Fe	215	7.9
Mg	44	1.74
Mn	200	7.4
Mo	338	10.2
Nb	104	8.6
Ni	208	8.9
Ti	106	4.5
V	127	6.2
W	393	19.3
Zr	96	6.4

Equation (4.22) shows that to minimize the mass m, we must maximize E/ρ. Therefore, the parameter to maximize, E/ρ, is known as the figure of merit for this material selection exercise. Table 4.2 lists the elastic modulus and density of common metallic elements. It is left to the reader to show that cobalt gives the highest E/ρ ratio.* In general, when selecting materials for a given application, other factors come into play in addition to mechanical properties, such as cost, availability, environmental impact, etc.

---

* In spite of its favorable modulus-to-density ratio and high melting point, chromium is currently not used as a structural material because of ductility considerations. Chromium has a BCC structure. At room temperature, few slip systems are activated. As a result, chromium is relatively brittle under ambient conditions. It would be quite exciting if someone could figure out a way to ductilize chromium.

4.11 BIOMEDICAL CONSIDERATIONS

Different materials (metals, polymers, and ceramics) are used in prosthetic devices. Durability is clearly of primary concern. Since body fluid is electrically conducting, materials placed inside the human body must be corrosion resistant because corrosion can generate products that may cause failure or undesirable immune response. As far as corrosion is concerned, polymers and ceramics are better materials. There are certain metallic alloys such as stainless steels, Co–Cr–Mo, and Ti alloys used in prosthetic devices with acceptably low corrosion rates. We will discuss corrosion and its prevention in a later chapter.

The other factor affecting durability is the mechanical properties of the implant material, which include wear resistance, elastic modulus, tensile strength, and toughness. In applications involving sliding contacts (e.g., artificial heart valves, knee, and hip joints), the material should have excellent wear resistance. In load-bearing implants such as the tibia or femur, the material should ideally have the same elastic modulus as the natural bone to simulate the body's compliance. For example, the average elastic modulus of the femur is around 15 GPa. The elastic modulus of polymers such as high-density polyethylene is about 1 GPa, while that of Ti-6Al-4V (i.e., titanium with 6 w/o aluminum and 4 w/o vanadium) is about 120 GPa. Therefore, an appropriate combination of metals and polymers can produce an interface that mimics the compliance of the human body.

The tensile strength of most implant materials exceeds that needed for regular daily use. Fracture toughness is another matter. In spite of the advantages of ceramics (excellent wear and corrosion resistance), ceramic materials generally have much lower fracture toughness than metallic alloys. Momentary impacts at high stress can produce cracks in ceramics that can have disastrous consequences. An obvious area of materials research is directed toward the science and technology of improving the fracture toughness of ceramic materials. We will talk more about this in the chapter on ceramics.

PROBLEMS

1. Given a wire of initial length L_o and an initial cross section A_o, a load F is applied to stretch this wire. Note that as the wire is stretched, the cross-section area will decrease because of the Poisson effect. Given that the wire material has a Young's modulus E, Poisson ratio v, and electrical resistivity ρ, and assuming completely elastic behavior, show that the resistance change ΔR is related to strain ε ($\varepsilon \ll 1$) according to the following. This is the basis of strain gauges:

$$\Delta R = \frac{\rho L_o}{A_o} \varepsilon (1 + 2v).$$

   (Hint: You need to determine the physical dimensions of the wire with an elastic strain of ε.)

2. a. For an iron bar of length 10 cm and cross-section area of 1 cm², calculate the force required in newtons to produce an elongation of 0.1 mm, assuming elastic deformation and a Young's modulus of 200 GPa. Be careful about units.

   b. For the same iron bar, we first apply stress of 400 MPa. This produces a total strain of 0.3%. The stress is then reduced to zero. The material sustains a permanent strain because the elastic limit is exceeded. Calculate the value of this permanent strain.

   c. The same iron bar has surface cracks 2 mm long. The relevant fracture toughness is 20 MPa-m$^{1/2}$ (2×10^7 Pa-m$^{1/2}$). Assume the geometric factor Y to be 1.0. What is the maximum stress this iron bar can withstand before fracture?

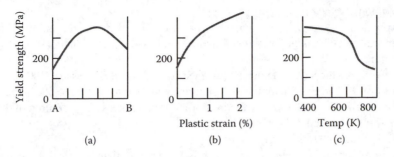

FIGURE 4.28 Dependence of yield strength on composition, plastic strain, and temperature.

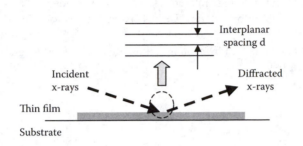

FIGURE 4.29 X-ray diffraction from a thin film.

3. Figure 4.28(a) is a plot of the yield strength for a substitutional alloy made of elements A and B as a function of composition. Figure 4.28(b) shows the variation of the yield strength of A versus plastic strain (cold working). Figure 4.28(c) shows the recovery of the yield strength of A after plastic straining to 1% versus annealing temperature. Based on this information, answer the following questions:
   a. What is the composition of the alloy giving the highest yield strength? State the percentage of A or B in the alloy. Be careful.
   b. State three different materials that can give a yield strength of 300 MPa.
   c. For material A after plastic straining to 1%, determine the highest temperature at which material A can be used, assuming that the acceptable yield strength is 300 MPa.
4. Imperfections such as dislocations can increase the strength of materials. Discuss two adverse effects on mechanical performance when the dislocation density in a material continues increasing.
5. Consider a thin film deposited on a substrate. X-ray diffraction is used to measure the interplanar spacing d as shown in Figure 4.29. Separate measurements show that the thin film is under biaxial compressive stress (i.e., parallel to the film surface). X-ray measurements, however, show that the interplanar spacing d is larger than the nominal (unstressed) value for this material. Explain the apparent discrepancy. (Hint: Think about the Poisson effect.)
6. Consider two adjacent, 1-mm long single crystal grains of iron to which a load of 2 N is applied (Figure 4.30). The Young's modulus along the [111] direction is 280 GPa, while that along the [100] direction is 125 GPa. The cross-section area for each grain is 1×1 mm^2. Assuming that the stress is uniform across the two grains, calculate the elongation of each grain. You will find that it is different for the two grains. Determine the shear strain.

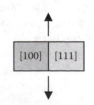

FIGURE 4.30 Deformation of a bicrystal.

7. Stress intensity factor:
   a. Explain in a sentence or two why the stress intensity factor is sometimes more important than yield strength or ultimate tensile strength of a material in dealing with fracture.
   b. A given mechanical structure is designed to withstand a stress of 200 MPa. Surface flaws on the order of 1 mm are present and must be tolerated. What is the minimum mode I critical stress intensity factor or fracture toughness (K_{Ic}) required of the material used in this structure?
8. Grain boundary effects:
   a. At room temperature, fine-grained metals are stronger than their single-crystal counterparts. State all the reasons why this is so.
   b. On the other hand, at high temperatures, exactly the opposite occurs because of atomic diffusion along grain boundaries. Discuss two methods to minimize or eliminate this problem.
9. When a material is stretched from an initial length L_o to a final length L_f, the engineering strain ε is given by:

$$\varepsilon = \int_{L_o}^{L_f} \frac{dL}{L_o} = \frac{L_f - L_o}{L_o}.$$

On the other hand, true strain ε_T is defined as dL/L, where L is the specimen length at a given applied stress. Therefore, as the material is stretched from L_o to L_f, the true strain is given by:

$$\varepsilon_T = \int_{L_o}^{L_f} \frac{dL}{L}.$$

Show that:

$$\varepsilon_T = \int_{L_o}^{L_f} \frac{dL}{L} = \ln(1+\varepsilon).$$

11. For a given alloy, experiments show that the fatigue life in a certain testing configuration is 1×10^6 cycles at a plastic strain amplitude of 1×10^{-3}. Noting that the Coffin–Manson exponent β is typically between -0.5 and -1.0 and assuming validity of the Coffin–Manson relation, make a conservative estimate of the plastic strain amplitude necessary to have fatigue life of (a) 10^7 cycles; and (b) 10^5 cycles. (Hint: Since β varies between -0.5 and -1.0, "conservative" means choosing the appropriate β [-0.5 or -1.0] to give the smaller plastic strain amplitude for the required fatigue life.)
12. Strength versus fracture toughness:
   a. Go into the literature to obtain a plot of strength versus fracture toughness for at least one class of material (e.g., steels or aluminum alloys). Note the inverse relationship.
   b. According to the discussion in the text, at a given external stress, there is a relationship between the size of the plastic zone and the yield strength. If the yield strength is doubled, how will this affect the size of the plastic zone in one dimension?

13. Consider a sphere placed on top of a perfectly rigid flat surface. A normal load is applied to the sphere, causing an elastic deformation of the sphere δ, given by:

$$\delta = \left(\frac{9N^2}{16E^2R} \right)^{1/3},$$

where N is the normal load, E is the elastic modulus of the sphere, and R is radius of the sphere. It is given that the maximum stress σ_{max} at the contact between the deformed sphere and the rigid flat is equal to:

$$\sigma_{max} = \frac{N}{\pi R \delta}.$$

There are two requirements in choosing the type of material for the sphere: (1) It will support the load such that $\sigma_{max} < \beta$, where β is some constant value; and (2) its mass M, equal to $4/3\pi R^3\rho$, where ρ is the mass density, should be as small as possible.

a. Starting with the inequality $\sigma_{max} < \beta$, show that $R/E > C$, where C is some constant.

b. Expressing R in terms of mass M and density ρ, derive the inequality $M > 4/3\pi\rho(CE)^3$.

c. Therefore, in order to obtain the smallest mass M, deduce the quantity to maximize (figure of merit) involving E and ρ.

5 Phase Diagrams

5.1 ROCKET NOZZLES*

In the pioneering days of space exploration, an engineer was asked to select the best material for rocket nozzles. The material must withstand exhaust gas temperatures up to 3000°C, must be tough and strong, and resist abrasion (due to debris coming off the exhaust). These early rockets used hydrocarbon fuels. The engineer decided to use tungsten, which appeared to satisfy all the requirements: high melting temperature (3410°C), tough, strong, and abrasion resistant. On the actual test stand, the nozzle material melted. Measurements during firing showed that the temperature was 3000°C, well below the melting temperature of tungsten. How could this have happened?

Subsequent chemical analysis showed that the original tungsten was quite pure, with less than 0.1 weight percent (w/o) carbon, while the tungsten sample dripped to the ground during the firing test contained 1.5 w/o carbon. The tungsten nozzle apparently picked up some carbon impurity, most likely from hydrocarbon fuel exhaust. Could this explain the unexpected melting of tungsten?

This is where phase diagrams can be helpful in deciphering the problem faced by the engineer in this story. A phase is the part of the material having uniform structure and composition, which can be solid, liquid, or gas. A phase diagram is a graph showing what phases are present at a given temperature, pressure, and composition and their relative concentrations. For each phase, we can learn at what temperature it begins to solidify or melt. Since the presence and concentration of these phases control properties, phase diagrams are important in the design and processing of metallic alloys, ceramics, and polymers. In the preceding example, inspection of the tungsten–carbon phase diagram shows that the addition of 1.5 w/o carbon to tungsten reduces the melting point from 3410 to 2460°C. This is well below the exhaust gas temperature of 3000°C; the nozzle failed simply because the temperature exceeded the melting point of the tungsten–carbon alloy.

In a given material system, chemical elements or compounds are present. They are known as components. For example, in the tungsten–carbon system,** there are two components: tungsten and carbon; in a system containing ice, water, and water vapor, H_2O is the only component. In this chapter, we will explore phase diagrams for single-component and two-component systems.

5.2 PHASE DIAGRAM FOR A SINGLE-COMPONENT SYSTEM: GRAPHITE/DIAMOND

Figure 5.1 shows the equilibrium phase diagram for the graphite/diamond system. Under normal temperature and pressure, graphite is the stable phase. Near room temperature and at pressures greater than 14 kbar (14,000 atm), diamond is the stable phase. At 1000 K, diamond is the stable phase when the pressure exceeds 34 kbar. Therefore, one way to make diamond is to subject carbon to high temperatures and pressures, a process developed by General Electric to make synthetic diamonds in the early 1950s. In principle, diamond is not a stable phase under normal ambient conditions. However, its transformation to graphite, the stable phase, is so slow that diamond can be considered to last forever at room temperature.

---

* Based on an actual occurrence, Prof. John Hilliard told this story in his article in *Journal of Educational Modules for Materials Science and Engineering*, 1:173 (1979).
** One important compound derived from tungsten and carbon is tungsten carbide, routinely used in cutting tools.

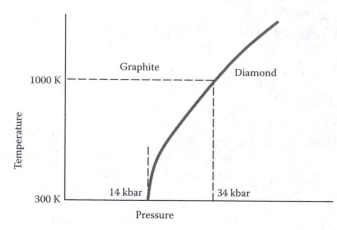

FIGURE 5.1 Phase diagram of carbon.

Figure 5.1 is an equilibrium phase diagram. Using nonequilibrium techniques, we can deviate from the equilibrium phase diagram and make diamond at low pressures. One such technique is plasma-enhanced chemical vapor deposition. By passing a mixture of 1% methane and 99% hydrogen (excited via some means of plasma discharge at 10^{-1}–10^{-3} torr) over substrates such as silicon held at 1000°C, we can readily grow polycrystalline diamond thin films. In this process, diamond and graphite form at the same time. Hydrogen atoms in the plasma discharge etch (remove) graphite faster than diamond, thereby resulting in the preferential growth of diamond.* By controlling the chemical composition (e.g., introduction of fluorine) and energetics of the plasma, it has been possible to grow diamond at temperatures as low as ~600°C, with typical rates on the order of microns per hour.

Question for discussion: Figure 5.1 shows that diamond is the stable phase at room temperature and pressures greater than 14 kbar. Can we make diamond from graphite at room temperature by squeezing graphite at greater than 14 kbar pressure? (Hint: The rate of transformation depends on the height of the activation barrier [see, for example, Figure 2.9], sometimes known as the kinetic barrier. For phase transformations involving carbon, the kinetic barrier is ~2–3 eV.)

5.3 PHASE DIAGRAM FOR A COMMON BINARY SYSTEM: NaCl + H_2O

At room temperature, a maximum of about 33 g of NaCl can be dissolved in 100 g of water. This is known as the solubility limit. If more sodium chloride is added to this saturated brine solution, the additional sodium chloride remains as a solid. Note that the solubility limit generally increases with temperature. Another important fact about this system is that the freezing point of water is reduced by the addition of sodium chloride. For example, a brine solution with 12 w/o of sodium chloride has a freezing point of –10°C instead of 0°C for pure water.** This is the basis of using

---

* William Eversole at Union Carbide Corporation made the first successful attempt to grow diamond films in the early 1950s by low-pressure chemical vapor deposition, using carbon-containing gases such as methane. The growth rate (a few nanometers per hour) was too low to be of practical interest at that time. Subsequently, John Angus at Case Western Reserve University reproduced these earlier results by Eversole and discovered that the introduction of hydrogen markedly increased the growth rate of diamond by several times. Optimization of this process was rigorously pursued by the Russian team led by Boris Deryagin and the Japanese team led by Nobuo Setaka. During this period, there was little diamond growth research in the United States and elsewhere. In 1984, the Japanese team demonstrated an unprecedented diamond growth rate of several microns per hour. This result ignited an explosive growth in diamond research worldwide.

** When water is free of dust or dirt, cooling as much as 10 to 20°C below the normal freezing point may be needed for ice to form. Dust or dirt acts as a nucleus and facilitates the formation of ice crystals.

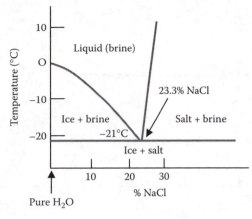

FIGURE 5.2 Phase diagram for the NaCl + H₂O system at ambient pressure.

salt to "melt" snow. The complete behavior of the NaCl + H₂O system is represented by the phase diagram shown in Figure 5.2 at an ambient pressure of 1 atm (760 torr).

Below –21°C, no liquid phase is present, and we have a mixture of two solids (ice and salt). This temperature is called the eutectic temperature, below which all phases are in the solid form. This means that we cannot use sodium chloride to melt snow below –21°C. The composition of water with 23.3 w/o NaCl is known as a eutectic.

Question for discussion: The preceding discussion suggests a method for desalination, a process of making drinking water from seawater. Discuss how this may work and the pros and cons of using this method for desalination. (Hint: Starting from room temperature, compare the energy it takes to cool to –21°C with that needed to boil water. The specific heat of water can be assumed to be 1 cal/g-°C. The latent heat of fusion is 80 cal/g, and the latent heat of vaporization is 540 cal/g. Once you have the ice and salt, how do you separate them?)

5.4   PHASE DIAGRAM FOR A BINARY ISOMORPHOUS SYSTEM: Ni + Cu

In this and subsequent discussion, we will assume that the ambient pressure is constant at 1 atm. For the Ni + Cu system, note that Ni and Cu have the same FCC structure (hence the term isomorphous) and similar atomic size and are completely soluble in each other.* Heat pure copper above 1085°C (melting point of copper) and then cool. Figure 5.3 shows the temperature of pure copper as a function of time. Note the flat portion of the curve (at 1085°C). This portion of the curve is flat because, as the liquid transforms to solid, latent heat of fusion is released, thus compensating for the heat loss during the cooling process and slowing the temperature decrease with time.

Figure 5.4 shows the same cooling curve for a Ni₅₀Cu₅₀ alloy.** Note the breaks at 1320 and 1260°C. The breaks in the cooling curve indicate that the Ni₅₀Cu₅₀ alloy contains a mixture of liquid and solid phases between 1320 and 1260°C. These temperature breaks can be experimentally determined as a function of composition, as shown in Figure 5.5. This is the phase diagram for the Cu–Ni system. The following definitions are used in denoting phase diagrams:

---

* The Ni–Cu alloy is known as a random alloy; that is, Ni and Cu atoms occupy the FCC lattice sites randomly. This can be contrasted with an ordered alloy — for example, Ni₃Al, an important FCC (properly known as L1₂) alloy used in turbine blades, in which Ni atoms sit at face centers and Al atoms at cube corners.
** Ni–Cu alloys are used in heat exchanger components and salt-water piping because of their good oxidation/corrosion resistance.

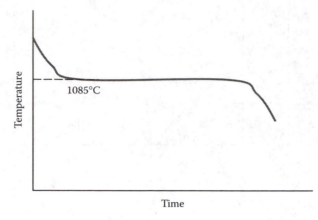

FIGURE 5.3 Cooling curve for pure copper.

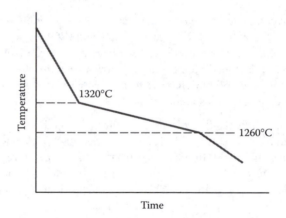

FIGURE 5.4 Cooling curve for a $Cu_{50}Ni_{50}$ alloy.

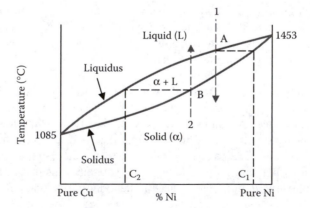

FIGURE 5.5 Phase diagram for the Cu–Ni system.

- The liquidus is defined as the curve above which all compositions are liquids.
- The solidus is defined as the curve below which all compositions are solids.
- Solid phases are usually labeled by Greek alphabet symbols, such as α- and β-phases, while the liquid phase is labeled L.

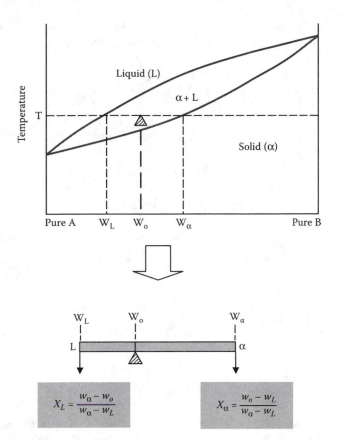

FIGURE 5.6 Schematic phase diagram for alloy AB to illustrate the lever rule.

In Figure 5.5, note the region enclosed between the liquidus and the solidus. This region consists of a solid phase (α) and the liquid phase (L) and is known as the two-phase region.

At a given alloy composition, we can use Figure 5.5 to follow the evolution of phases. For example, cooling from the liquid phase (L) along path 1, a solid starts to form at point A. The composition of the solid is given by C_1 and is different from that of the starting liquid. Similarly, warming from the solid phase (α) along path 2, liquid starts to form at point B. The composition of the liquid is given by C_2 and is different from that of the starting solid.

5.4.1 THE LEVER RULE

Many engineering alloys consist of two or more phases. Their physical and chemical properties depend on the percentages of phases present. How are the percentages of phases determined within the two-phase region? Specifically, based on the phase diagram shown in Figure 5.6, at a given composition and temperature, what fraction is the α-phase and what fraction is the L-phase in the two-phase region? This is done by the lever rule, derived as follows. Consider an alloy with weight fraction w_o of B at temperature T, as marked by the shaded triangle in the two-phase region (Figure 5.6). The following steps are involved:

- Draw a horizontal line through the shaded triangle so that it intersects the liquidus and solidus.
- Drop vertical lines from the two intercepts to the horizontal (composition) axis to obtain w_L, the weight fraction of B in the liquid (L) phase, and w_α, the weight fraction of B in the α-phase.

- Assume that we have 100 g of the alloy. The total amount of B present $= 100\, w_o$.
- Assume that the weight fraction of the α-phase $= X_\alpha$ and that of the L-phase $= X_L$ (note that $X_\alpha + X_L = 1$).
- Total amount of B in the α-phase $= 100\, X_\alpha w_\alpha$.
- Total amount of B in the L-phase $= 100\, X_L w_L$.
- Conservation of B implies that:

$$100\, w_o = 100\, X_\alpha w_\alpha + 100\, X_L w_L$$

$$= 100\, X_\alpha w_\alpha + 100\, (1 - X_\alpha)\, w_L$$

$$\text{i.e., } w_o = X_\alpha w_\alpha + (1 - X_\alpha)\, w_L$$

$$= X_\alpha\, (w_\alpha - w_L) + w_L.$$

Solving, we have:

$$X_\alpha = \frac{w_o - w_L}{w_\alpha - w_L}; \tag{5.1a}$$

$$X_L = \frac{w_\alpha - w_o}{w_\alpha - w_L}. \tag{5.1b}$$

Equation (5.1) allows us to calculate weight fractions of α- and L-phases. The sketch at the bottom of Figure 5.6 provides a memory aid for Equation (5.1) and shows the origin of the lever rule name.

EXAMPLE

From Figure 5.6, start with an alloy with 50 w/o of B. The temperature is increased to bring this alloy into the two-phase region. Values of w_L and w_α are 10 and 60 w/o, respectively. How much of this alloy is in the form of a liquid?

Solution

The easiest approach to a problem of this nature is to lay out all three values as follows:

$$10 \qquad\qquad\qquad\qquad 50 \quad 60,$$

where $w_L = 10$ w/o, $w_o = 50$ w/o, and $w_\alpha = 60$ w/o. From Equation (5.1b), we have:

$$X_L = \frac{w_\alpha - w_o}{w_\alpha - w_L} = \frac{60 - 50}{60 - 10}$$

$$= \frac{10}{50} = 0.2.$$

Therefore, at the temperature given, 20% of this alloy is present as a liquid.

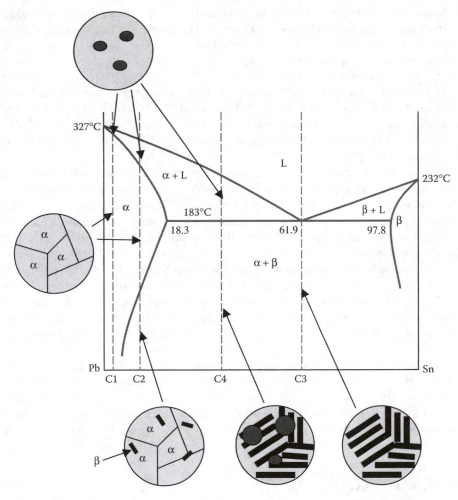

FIGURE 5.7 The Pb–Sn phase diagram.

5.5 BINARY EUTECTIC ALLOYS: MICROSTRUCTURE DEVELOPMENT

The sodium chloride + water system is an example of a binary eutectic system. Another techno-logically important binary eutectic system is the Pb–Sn system (Figure 5.7), which is used as a solder alloy in electrical circuits and microelectronics packaging.* Note the limited mutual solubility in this system. By examining Figure 5.7, we can immediately learn the following:

- Pure Pb melts at 327°C, and pure Sn melts at 232°C.
- Pb dissolves a maximum of 18.3 w/o Sn.
- Sn dissolves a maximum of 2.2 w/o Pb.
- The eutectic alloy contains 61.9 w/o Sn and melts at 183°C.

Raise the temperature of a lead–tin alloy of arbitrary composition to bring the alloy into the liquid phase. Then allow it to cool slowly from the melt. The resulting microstructure depends on the composition. Let us consider four different compositions as shown in Figure 5.7:

---

* Because of the concern about lead poisoning, there has been a strong push to phase out the use of lead in electronics since the 1990s. Replacement solder materials include indium, silver, and tin.

C1. As the alloy cools from the melt, the α-phase (mostly pure lead) precipitates. Because of the small temperature difference between the two-phase and single-phase regions, the α-phase grows readily to form a polycrystalline solid. The grain size depends on the cooling rate; slower cooling results in large grain size. Because of the low tin concentration, the resulting solid contains a single phase.

C2. As the cooling brings the alloy into the single-phase α-region, a polycrystalline solid forms with the α-phase only, similar to the case of C1. Further cooling below the line separating α- and α + β-phase fields (this is known as the *solvus* line) results in the precipitation of the β (tin-rich)-phase. The β-precipitates occur within the α-grains or on grain boundaries.

C3 (eutectic). As the eutectic melt cools below the eutectic temperature, solidification begins with the simultaneous formation of α (lead-rich)- and β (tin-rich)-phases. Since the original melt has a uniform composition, the transformation to the two solid phases with different compositions requires diffusion of lead and tin atoms. The resulting microstructure consists of alternating α- and β-layers (lamellae). The lamella thickness depends on the cooling rate; it gets thinner with faster cooling.

C4. When the cooling brings the alloy into the two-phase region, precipitation of the α-phase occurs, usually in the form of spheroidal grains. Below the eutectic temperature, the remaining liquid has to transform into a mixture of α- and β-phases. Similar to the preceding case, the solidification process requires simultaneous diffusion of lead and tin atoms, resulting in the formation of a lamella microstructure intermixed with the spheroidal α-grains.

As discussed earlier, mechanical properties can be controlled by alloying through solid solution strengthening. Here, we show that alloying also results in different microstructures such as precipitates, which in turn can have marked effects on mechanical properties (strength, fatigue life, ductility, and toughness). The cooling rate affects both the grain size and thickness of individual lamellae. The ability to control microstructure through alloying and heat treatment is one of the many tools in a materials scientist's bag of tricks.

5.6 ZONE REFINING

One technique to obtain high-purity materials is zone refining. This can be understood using phase diagrams. Consider a silver sample with small amounts of copper impurities. To purify silver, we first pass the silver sample through a zone heater at a temperature slightly below the melting point of pure silver. We can observe the formation of a liquid film on the sample surface. This liquid is enriched with copper (i.e., the liquid contains a higher concentration of copper impurity than the solid sample). Therefore, copper is preferentially extracted from the sample. The liquid is then discarded. Repetition of this process produces an increasingly pure silver sample.

The basic principle why this zone-refining technique works can be understood by examining the section of the Ag–Cu phase diagram near the pure Ag side (Figure 5.8). Consider that the original silver sample weighs 100 g containing 5 w/o copper. Assume that we heat the sample to 900°C. At this temperature, we have two phases: liquid (*L*) containing 9.4% Cu and solid (α) containing 3% Cu. Note that the liquid is enriched in Cu. Using the notation of Equation (5.1) for the lever rule, the weight fraction of the solid phase X_α is given by:

$$X_\alpha = \frac{w_o - w_L}{w_\alpha - w_L} = \frac{5 - 9.4}{3 - 9.4}$$

$$= \frac{4.4}{6.4} = 0.69.$$

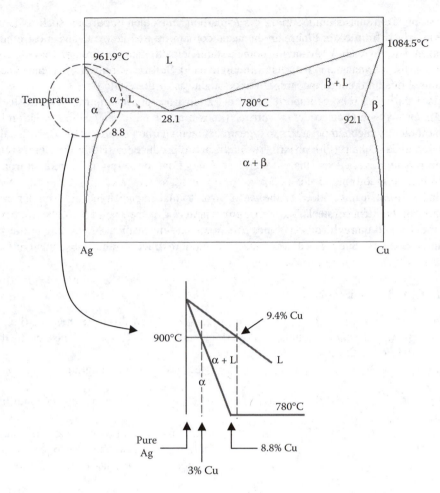

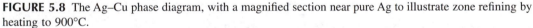

FIGURE 5.8 The Ag–Cu phase diagram, with a magnified section near pure Ag to illustrate zone refining by heating to 900°C.

Therefore, at 900°C, the weight fraction of the solid phase (α) is 69% (i.e., the weight of the solid phase is 69 g). The amount of Cu in the solid phase at this temperature is:

$$= 100 \, X_{\alpha} w_{\alpha}$$

$$= 100 \times 0.69 \times 0.03$$

$$= 2.1 \text{ g,}$$

where the concentration of Cu in the solid phase w_{α} is 3%. This copper concentration is lower than that of the original alloy (5%).

5.7 APPLICATION OF PHASE DIAGRAMS IN MAKING STEELS

5.7.1 PRODUCTION OF IRON AND STEELS

Iron ores are mainly in the form of oxides. The first step in making steel is to reduce iron oxide to metallic iron by heating a mixture of coke (mostly carbon) and the iron ore. The resulting iron is known as pig iron and contains up to 4 w/o carbon and other impurities such as sulfur and phosphorus derived from coke. Plain carbon steels contain up to 1.2 w/o C and most other steels contain less than 0.5 w/o C. While having some carbon helps to increase strength, having too much carbon reduces the toughness and makes it difficult to weld. Sulfur and phosphorus tend to segregate to grain boundaries and reduce the strength and toughness of these alloys.

To reduce carbon and other impurity content in pig iron, pig iron is mixed with steel scrap, lime (CaO), and oxygen in an oxidation furnace (known as the basic oxygen or oxidation furnace, BOF). At sufficiently high temperatures, oxygen reacts with carbon to produce CO, with sulfur and CaO to produce $CaSO_4$, and with phosphorus and CaO to produce $Ca_3(PO_4)_2$. Calcium sulfate and calcium phosphate are less dense than molten iron. They float to the top of the molten iron known as slag and can be readily removed.

Alloying elements can be added to the molten iron to make specialty steels (smaller grain size for higher strength, stainless steels, etc.). The molten steel can then be cast into slabs. Properties can be further refined through additional heat treatment and mechanical processing such as rolling or forming operations. Such processing affects the microstructure and subsequent mechanical properties.

5.7.2 Fe–Fe₃C Phase Diagram

Plain carbon steels contain up to 1.2% C, with 0.25–1% Mn for grain refinement (i.e., having smaller grain size) and small amounts of impurities such as Si, P, and S. The key to the steel technology is the Fe–Fe$_3$C phase diagram (Figure 5.9).

Let us start by describing the three principal phases shown in this diagram:

- Ferrite (α) is an interstitial solution of C in BCC iron;* maximum carbon solubility is 0.022 w/o at 727°C.
- Austenite (γ) is an interstitial solution of C in FCC iron. The maximum carbon solubility is 2.14 w/o at 1147°C. At 727°C, the carbon solubility in austenite is 0.76 w/o — much higher than that in ferrite. This higher solubility is due to the larger size of interstitial sites in an FCC lattice.
- Cementite ($\bar{c}$) or iron carbide (Fe$_3$C) is hard and brittle and contains 6.67 w/o C.**

Table 5.1 shows a summary of terminologies used to describe different iron-based phases. The phase transformation from BCC ferrite to FCC austenite involves a volume decrease (recall that an FCC unit cell is more tightly packed than a BCC one). This volume decrease is often demonstrated by heating up a piano wire (made of low-carbon steel) fixed between two posts. As the temperature increases, thermal expansion causes the piano wire to sag. As the ferrite–austenite transition temperature is crossed, the volume shrinkage results in a shorter wire and reduces the wire sagging momentarily (Figure 5.10). With further temperature increase, normal thermal expansion resumes and increases the wire sagging.

Fe containing exactly 0.76 w/o C at 727°C is known as a eutectoid or eutectoid steel. Hypereutectoid steels contain more than 0.76 w/o C, while hypoeutectoid steels contain less than 0.76

---

* The term "ferrite" also refers to a class of iron-based oxides used in magnetic tapes.
** An old Chinese folk tale tells of a couple well known for their swordsmithing skills. In anticipation of the new emperor's inauguration, the imperial court commissioned this couple to make the strongest sword as a gift to the emperor. No matter how hard they tried, the couple managed only to make swords as strong as putty. With the deadline fast approaching, rather than face the wrath of the imperial court and certain death, the couple sacrificed themselves by jumping into the pot of molten iron that was used to make the sword. Miraculously, when their apprentices made the sword from this molten iron, the sword was the strongest they had ever made, and they thanked the gods for their success. Regardless of divine intervention, it appears that increasing the carbon content through the couple's selfless act may have played a critical role in the eventual success of making stronger steels.

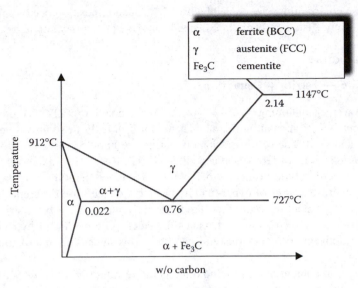

FIGURE 5.9 The Fe–Fe$_3$C phase diagram.

TABLE 5.1
Description of Different Iron-Based Microstructures

Name	Description
Ferrite (α)	Fe + <0.022 w/o C, body-centered cubic structure
Austenite (γ)[a]	Fe + C, face-centered cubic structure
Cementite ($\bar{c}$)	Fe$_3$C
Eutectoid steel	Fe + 0.76 w/o C
Hypoeutectoid steel	Fe + <0.76 w/o C
Hypereutectoid steel	Fe + >0.76 w/o C
Pearlite	Ferrite + cementite, obtained by cooling austenite to between 540 and 727°C, lamella structure
Upper bainite	Ferrite + cementite, obtained by cooling austenite to between 300 and 540°C
Lower bainite	Ferrite + cementite, obtained by cooling austenite to between 200 and 300°C
Martensite[b]	Fe + supersaturated C in solution, obtained by quenching austenite to room temperature, body-centered tetragonal structure
Ferritic steels	Fe + C alloys with the body-centered cubic structure
Austenitic steels	Fe + C alloys with the face-centered cubic structure
Pearlitic steels	Equilibrium Fe + C alloys with the pearlite microstructure

[a] Also used to designate the high-temperature phase of a shape memory alloy.
[b] Also used to designate the low-temperature phase of a shape memory alloy.

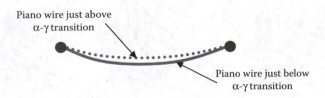

FIGURE 5.10 Piano wire experiment, showing the shrinkage as the wire is heated just above the α–γ transformation temperature.

w/o C. When these compositions are cooled below 727°C, decomposition occurs to form a mixture of ferrite and cementite. The microstructure depends on the exact composition, as discussed next.

5.7.3 Microstructure

5.7.3.1 Austenite → Ferrite + Cementite

Take a plain carbon steel containing 0.76% C (eutectoid) and heat it to >727°C to produce austenite. This heat treatment is known as austenitizing. Then cool it quickly to some temperature between 727 and ~540°C and hold. The result is a lamellar structure known as pearlite (Figure 5.11) that is made of alternating layers of ferrite and cementite (Fe_3C). Recall the analogous lamellar microstructure when a eutectic solution is cooled below the eutectic temperature. It is important to note, however, that the microstructures of hyper- and hypoeutectoid steels are different (Figure 5.11). In hypereutectoid steels, cementite forms first, predominantly along grain boundaries, followed by pearlite. This creates a microstructure that makes the steel more brittle. In hypoeutectoid steels, ferrite forms first along grain boundaries and pearlite forms later. This microstructure gives the steel higher toughness.

For a given carbon content and in the hold temperature range of 727 to ~540°C, the lamella thickness in the pearlite structure decreases with decreasing temperature. Why? Recall that the formation of pearlite requires the diffusion of carbon atoms. The lower the temperature is, the shorter is the diffusion distance and hence the thinner is the lamellae.

5.7.3.2 Bainite

What would happen if we further reduce the hold temperature? Diffusion is sufficiently slow that the uniform lamellar structure of pearlite does not form. Instead, at temperatures between ~300 and 540°C, we observe the formation of irregular parallel strips of ferrite separated by elongated particles of cementite. This microstructure is known as upper bainite. Between ~200 and 300°C, the microstructure (lower bainite) consists of thin plates or needles of ferrite embedded with fine cementite particles.

Question for discussion: If we heat a steel sample with the pearlite or bainite microstructure to 700°C and hold it there for a long time (24 h), discuss what may happen to the shape of the Fe_3C phase. (Hint: At thermal equilibrium, a system wants to minimize its free energy. What is the particle shape that gives the minimum surface free energy?)

5.7.3.3 Martensite

The preceding microstructures are formed under relatively slow cooling conditions. If the cooling is performed rapidly (quenching) — for example, by dropping the sample in oil or water — carbon diffusion cannot occur fast enough to form the pearlite or bainite structure. A nonequilibrium needle-shaped microstructure known as martensite will result. It is a metastable phase of BCC iron supersaturated with carbon atoms in interstitial solution. The carbon atoms sit at interstitial sites

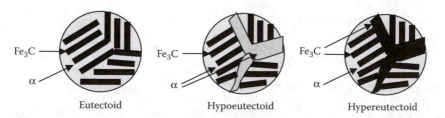

| Eutectoid | Hypoeutectoid | Hypereutectoid |

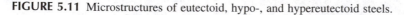

FIGURE 5.11 Microstructures of eutectoid, hypo-, and hypereutectoid steels.

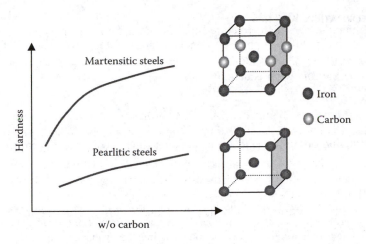

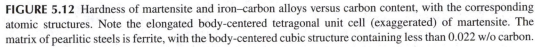

FIGURE 5.12 Hardness of martensite and iron–carbon alloys versus carbon content, with the corresponding atomic structures. Note the elongated body-centered tetragonal unit cell (exaggerated) of martensite. The matrix of pearlitic steels is ferrite, with the body-centered cubic structure containing less than 0.022 w/o carbon.

on the side of the iron unit cell. This causes the unit cell to stretch slightly along one direction, resulting in a body-centered tetragonal (BCT) structure (Figure 5.12). Unlike the transformation of austenite to ferrite and cementite, no diffusion is required.* Because of the presence of a supersaturated solution of carbon and the existence of a noncubic structure, slip does not occur easily in the martensite phase. As a result, martensite is much harder and more brittle than the corresponding iron–carbon alloy (pearlitic steels) with the same carbon concentration obtained by slow cooling (Figure 5.12).

While the high hardness of martensite makes it strong against and resistant to abrasion, its brittleness can be a problem. This can be solved by a heat treatment process known as tempering. When heated to a temperature below the eutectoid temperature (727°C), transformation to a mixture of ferrite and cementite (the equilibrium phases) occurs. This alloy is known as tempered martensite; it consists of coarser cementite precipitates and has lower strength but higher toughness than martensite. By controlling the tempering conditions, steels with an optimum combination of strength and toughness can be produced. In other systems, such tempering or annealing treatment may result in strength increase.**

Question for discussion: As shown in the phase diagram, steels can be made with ferrite or austenite as the primary phase. Discuss the advantages of austenitic over ferritic steels. (Hint: What is the difference in structure between these two steels? How is this difference reflected in toughness and its dependence on temperature?)

---

* The transformation of austenite (FCC) to martensite (BCT) involves lattice distortions that require atomic displacements on the order of 0.1 nm only. As a result, the martensitic transformation is also called a displacive or diffusionless transformation.

** The concept of using quenching and annealing to obtain specific microstructure and strength is key to the design of aluminum alloys used in rivets, which are used extensively as fasteners in aircraft. During the fastening process, the aluminum alloy should ideally be soft and ductile so that it will fit snugly. However, if the alloy is soft, it may not have the strength to hold things together. This dilemma is solved by the following scheme. The alloy is aluminum with 4 w/o copper. All copper atoms are first brought into a solid solution by raising the temperature above 500°C, followed by quenching in ice water. The quenching retains the solid solution of copper in aluminum, which is still relatively soft. The softness is retained as long as the alloy is maintained below room temperature. When it is ready to be used in the riveting operation, the alloy is taken out of cold storage and used immediately. At room temperature, precipitation of small and much stronger intermetallic alloy particles (Al_3Cu_2) occurs. These precipitates interfere with dislocation movements and make the alloy stronger. This is known as age- or precipitation hardening. This is an excellent example illustrating one of the fundamental principles presented earlier in this book: *Properties depend on synthesis or processing techniques.*

5.7.4 TRANSFORMATION KINETICS

Transformation between different phases occurs at a certain rate — for example, from austenite to ferrite + cementite ($\gamma \to \alpha$ + carbide). We can describe these transformation processes, such as those described by the following equation:

$$x = \exp(-kt^n), \tag{5.2}$$

where x is the fraction of the phase that remains after time t, with k and n constants.

EXAMPLE

The t^n term in Equation (5.2) is somewhat unusual, compared with other rate processes we may have encountered before — for example, in radioactive decay, $x = \exp(-kt)$. Consider the case when the transformation rate is proportional to concentration and time. Show that $x = \exp(-at^2)$, where a is some constant.

Solution

Given that the rate of transformation is proportional to concentration and time, we can write:

$$\frac{dx}{dt} = -kxt.$$

Transposing, we have:

$$\frac{dx}{x} = -ktdt.$$

Solving yields:

$$x = \exp\left(-\frac{1}{2}kt^2\right).$$

In the transformation of $\gamma \to \alpha$ + carbide, the temperature T must be below a critical temperature T_{cr} (727°C) for the transformation to occur. The transformation rate depends on two factors: how far the temperature is from thermal equilibrium ΔT ($= T_{cr} - T$) and diffusion rate of atoms (T). When T is slightly below T_{cr}, the transformation is slow because it is close to equilibrium. As shown in Figure 5.13, it takes a long time to transform a certain percentage (99% in this case) of the γ-phase to α + carbide just below T_{cr}.

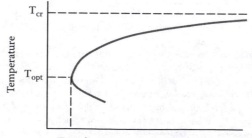

FIGURE 5.13 Schematic diagram showing the dependence of transformation rate (as measured by the time for 99% of the material to transform) on temperature T below the critical temperature T_{cr}.

As we move further from equilibrium (lower temperature), the transformation rate increases until it reaches an optimum temperature T_{opt} at which it takes the shortest time for the transformation to occur. Below this optimum temperature, the diffusion rate of atoms becomes rate limiting, so the transformation time increases. The curve shown in Figure 5.13 is known as the TTT (temperature–time–transformation) diagram.

5.7.5 ALLOYING ELEMENTS

Alloying elements are often added to steels to improve chemical and physical properties.

- An alloying element can be incorporated to form a solid solution with the steel matrix. The result is solid-solution strengthening, usually without much loss of ductility.
- Elements such as chromium and tungsten form hard carbides, resulting in an increase in strength.
- Some elements stabilize the ferrite (BCC) phase relative to the austenite phase and thus increase the ferrite–austenite transformation temperature. Examples include chromium, tungsten, and silicon. On the other hand, elements such as nickel and manganese stabilize the austenite phase and hence decrease the transformation temperature. In fact, when sufficient concentrations of nickel are added, the ferrite phase may be completely eliminated.*
- The addition of greater than 12% chromium to steels results in the formation of a protective oxide for the underlying matrix, thus imparting the steels with oxidation and corrosion resistance. This is the basis of stainless steels.
- Some elements such as nickel and silicon destabilize cementite and are usually not introduced to high-carbon steels.
- The addition of elements by themselves or via reactions with the matrix to form fine precipitates increases alloy strength and, in some cases, toughness.
- Increased alloy strength and fracture toughness can be achieved by grain refinement. A good example is manganese.

Question for discussion: Strictly speaking, Figure 5.9 is not an equilibrium phase diagram because graphite, instead of Fe3C, is the equilibrium phase. Under equilibrium conditions, carbon introduced into iron in excess of the solubility limit precipitates as graphite. Discuss or speculate about ways to promote the precipitation of carbon as diamond rather than graphite without resorting to high pressures and temperatures.

5.7.6 AISI-SAE NAMING CONVENTIONS

Plain carbon steels are given the 10XX designation, where XX is equal to the weight percent of carbon multiplied by 100. Alloy steels are given the WWXX designation, where WW represents the type of alloy addition and XX represents the weight percent of carbon multiplied by 100:

1040: plain carbon steel typically used in crankshafts, nuts, and bolts, with 0.40 w/o carbon and typically 0.45 w/o Mn

4340: alloy steel used in bushings and aircraft struts and tubings, containing 1.71 Ni, 0.77 Cr, 0.3 Mo, 0.75 Mn, and 0.40 C

8620: alloy steel used in gears, containing 0.55 Ni, 0.50 Cr, 0.2 Mo, and 0.20 C

52100: alloy steel used in bearings, containing 1.45 w/o of Cr and 1.0 w/o of C

---

* The effect of alloying elements on the ferrite–austenite transition temperature can be determined via careful thermodynamic calculations. However, we can make a plausible argument to make some predictions. For example, the addition of FCC elements such as Ni stabilizes the FCC austenite phase and hence lowers the transition temperature to austenite; the addition of BCC elements such as Cr stabilizes the BCC ferrite phase and hence raises the transition temperature to austenite.

5.8 SHAPE MEMORY ALLOYS

Shape memory alloys have two unique properties: the shape memory effect and super- or pseudo-elasticity. These properties were first observed in 1938. The most commonly used shape memory alloys are NiTi, CuZnAl, and CuNiAl. These unusual properties appear due to the transformation from one solid phase with one atomic structure to another phase with a different atomic structure. The transformation is driven not only by temperature, but also by stress. The two phases involved in this transformation are martensite (low-temperature phase) and austenite (high-temperature phase).

Martensite is the stable phase at low temperatures. It has a twinned structure. When it is heated above a critical temperature, it transforms into the high-temperature stable phase known as austenite (Figure 5.14), which has a cubic structure. A classic way to demonstrate the shape memory effect is as follows:

1. Start with an alloy that has a specific shape above the austenite transformation temperature. The alloy has the cubic (austenite) structure.
2. Cool the alloy to acquire the martensite structure.
3. Deform it (e.g., by bending or twisting).
4. Warm it above the austenite transformation temperature. The alloy returns to the austenite structure and acquires the original shape — hence the term "shape memory alloy."

The second unique property, super- or pseudoelasticity, arises from the dependence of phase stability on stress. Let us first define two quantities: M_{50} and A_{50}. As discussed in the section on the kinetics of phase transformation, it takes a certain amount of time for phases to transform at a given temperature. In this discussion, we define M_{50} as the temperature at which 50% of the phase has

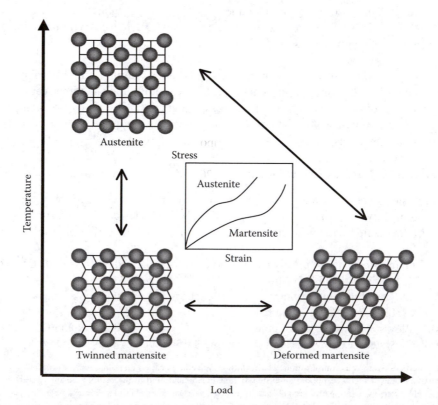

FIGURE 5.14 Transformation among three different structures: martensite, deformed martensite, and austenite.

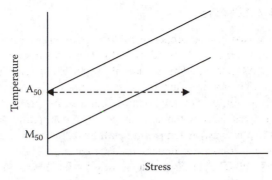

FIGURE 5.15 Dependence of austenite and martensite transformation temperature on applied stress.

transformed into martensite during the time of the experiment. Likewise, we define A_{50} as the temperature at which 50% of the phase has transformed into austenite during the time of the experiment.

It turns out that stress affects the transformation temperatures of a given shape memory alloy (Figure 5.15). Starting with austenite, apply stress to the alloy at temperatures greater than or equal to A_{50}. When the stress exceeds a certain value, the alloy transforms into martensite as shown.* For NiTi, the martensite phase is softer and can sustain a larger strain. When the stress is removed, the alloy returns to the austenite structure, recovering the original shape — hence the term pseudoelasticity.

Question for discussion: Applications of shape memory alloys include thermostats, wire frames for eyeglasses, and vascular stents. There have been suggestions of incorporating shape memory alloys in the design of self-healing materials (e.g., in closing cracks). How can this be done?

5.9 PHASE TRANSFORMATION IN BIOLOGICAL SYSTEMS: DENATURATION OF PROTEINS

5.9.1 BACKGROUND

DNA carries genetic information, and the execution of a given task directed by that information is through proteins. Each cell contains thousands of proteins serving a wide range of functions, such as enzymes, hormones, defense against infections, structural integrity of cells, and transport of small molecules through cell membranes. The importance of proteins can be recognized from the Greek root *proteios*, which means "of first rank."

The basic building blocks of proteins are the 20 amino acids. Each consists of a carbon atom (the α-carbon) bonded to a hydrogen atom, an amino group (NH_3^+), a carboxyl group (COO^-), and a side chain (R), as shown in Figure 5.16. A protein molecule consists of a unique sequence of such amino acids.** The linkage between different amino acids is from the carboxyl group (COO^-)

---

* When we talk about stress-induced transformation, the natural question is: Does it matter whether the stress is tensile or compressive? Two things accompany martensitic transformation: shear and volume change. Shear can result from tensile or compressive stress. If volume expansion occurs during the austenite–martensite transformation (as is the case with NiTi and steels), tensile stress will be favored. The net result is that tensile or compressive stress can cause martensitic transformation, although there is some degree of asymmetry, depending on the sign and magnitude of volume change occurring during the transformation.

** For those who took biology years ago, here is a refresher. The synthesis of a protein molecule from DNA goes through the following sequence. First, the double helix DNA unzips. A template of an appropriate portion of the unzipped DNA is synthesized, a process known as transcription. The resulting molecule is called messenger-RNA (m-RNA). Transfer RNA (t-RNA) molecules bind to the appropriate amino acids on one side and bind to the m-RNA on the other with the correct nucleotide matching, thus creating the amino acid sequence for a given protein. This process is known as translation. Molecular biology experiments show that each amino acid is coded by three nucleotides. Since there are four nucleotides, there are 64 possibilities ($4^3 = 64$), of which 61 code for the 20 amino acids and 3 code for termination of protein synthesis.

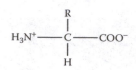

FIGURE 5.16 Structure of an amino acid.

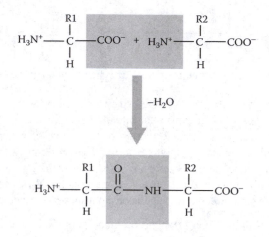

FIGURE 5.17 Formation of a peptide bond.

of one amino acid to the amino group (NH_3^+) of another, via the elimination of water, as shown in Figure 5.17. This linkage is known as a peptide bond.

5.9.2 PROTEIN CONFORMATION

A protein molecule is more than a spaghetti string of amino acids. The CO group in one peptide bond interacts with the NH group in another peptide bond through hydrogen bonding. These interactions cause the protein chain to form helices (known as α-helices) and pleated sheets (known as β-sheets). Together with interactions between side groups, the protein chain folds and adopts a distinct three-dimensional shape or conformation that endows the molecule with specific functions. This unique shape can be destroyed by energetic excitation (e.g., heating) or by changing the chemical environment (e.g., changing the pH) to overcome these interactions. This process is known as denaturation, which may or may not be reversible. As schematically illustrated in Figure 5.18, we can consider the protein molecule in its native coiled state as one phase and the linear chain as another phase. The transformation is driven by temperature change, analogous to the ferrite–austenite or ice–liquid water transformation.

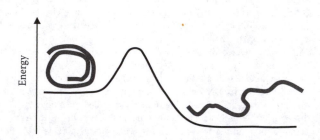

FIGURE 5.18 Denaturation of a protein molecule.

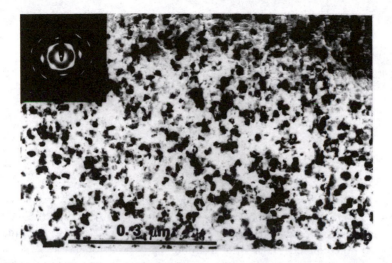

FIGURE 5.19 Transmission electron micrograph of a $Ag_{0.33}Mo_{0.67}$ film. The dark regions are due to Mo. The inset is an electron diffraction pattern. The relatively large diffraction spots indicate that the material is nanocrystalline. (Reprinted with permission from Dr. Y. T. Cheng of General Motors.)

5.10 APPLICATION OF PHASE DIAGRAMS IN MAKING NANOCRYSTALLINE MATERIALS

According to the Hall–Petch relationship, the strength of a material depends on some inverse power of the grain size. Therefore, materials with grain size in the nanometer scale should be very strong, as has been shown experimentally.

One way to make nanocrystalline materials is by codeposition of two species that have little or no mutual solubility. The degree of immiscibility for any two elements can be determined from phase diagrams. The resulting film usually consists of nanocrystalline grains of one phase, separated from one another by a continuous matrix of the second phase. One example is the Ag–Mo system.* The lack of mutual solubility in this system implies that the Ag–Ag and Mo–Mo bond strengths are greater than that for Ag–Mo. This drives the phase separation, resulting in high nucleation rates and nanometer grain size (Figure 5.19).

The two components can also be compounds. For example, when titanium nitride and silicon nitride are codeposited at an appropriate ratio, the resulting film consists of titanium nitride nanocrystalline grains separated from one another by amorphous silicon nitride. The nanocrystalline structure is retained even after annealing in an inert environment (vacuum or argon) for an hour at 1000°C. Because of the nanometer grain size, these films have enhanced mechanical properties, with hardness values in the 40- to 60-GPa regime. This can be compared with 20–25 GPa for "normal" titanium nitride and 15–20 GPa for amorphous silicon nitride. Therefore, this provides a novel strategy to synthesize superhard wear-protective coatings.

5.11 PHASE DIAGRAMS FOR DENTISTRY

Dentists have been using mercury amalgams for a long time to fill cavities (some say as far back as the 16th century — it is a fair assumption that no one worried about mercury poisoning in those

---

* Several factors determine the degree of mutual solubility: atomic size, crystal structure, electronegativity, and valence, as stated in Chapter 2. For the Ag–Mo system, Ag has a face-centered cubic structure, while Mo is BCC. The atomic radius of Ag is 0.144 nm versus 0.136 nm for Mo. The electronegativity of Ag is 1.93 versus 2.16 for Mo. The valence of Ag is one compared with the multiple valence states for Mo.

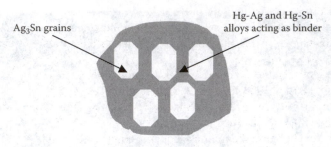

FIGURE 5.20 Schematic microstructure of a low-copper dental amalgam.

days). In the mid-1800s, silver with 5–10 w/o of copper, similar to the composition of silver coins used then, was tried and abandoned because of its tendency to discolor and fracture.

Dental amalgam as delivered to the dentist has two components: a powder of primarily silver–tin alloy (Ag$_3$Sn) and liquid mercury. Just prior to filling, appropriate quantities of these two components are mixed together vigorously (the process is known as trituration). During the mixing process, mercury reacts with Ag$_3$Sn to form low-melting-point Ag$_2$Hg$_3$ and Sn$_7$Hg intermetallic compounds. The initially formed mercury-containing mixture (amalgam) is soft and can be placed and shaped into the cavity with relative ease before subsequent cooling and hardening. The mercury intermetallic compounds can be thought of as the glue between Ag$_3$Sn grains (Figure 5.20).

The addition of copper up to 30 w/o appears to improve the marginal integrity of the filling. In spite of the relative softness of the individual components, dental amalgams of different formulations are sufficiently wear resistant. However, in addition to mercury poisoning concerns, corrosion is always a problem with metallic fillings, leading to loosening and eventual filling loss. Furthermore, the silvery/dark color of metallic fillings is not the most appealing from an aesthetic point of view; new corrosion-resistant and more aesthetically appealing materials are needed.

PROBLEMS

1. Consider a section of the tungsten–carbon phase diagram, as shown in Figure 5.21. What is the maximum carbon content in tungsten that can yield an alloy with melting point greater than 3000°C?

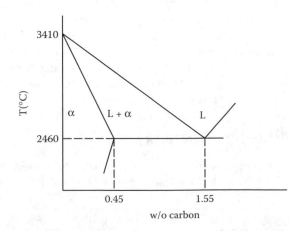

FIGURE 5.21 A section of the tungsten–carbon phase diagram.

2. Refer to the Pb–Sn phase diagram in the text.
   a. State the composition of the eutectic alloy and the eutectic temperature.
   b. What will happen to the eutectic alloy when the temperature is raised above the eutectic temperature?
   c. Cool 100 g of the eutectic alloy to slightly below the eutectic temperature. Calculate the amount of α- and β-phase at thermal equilibrium.
   d. Take a Pb–Sn alloy with 25 w/o Sn. As we heat the alloy from room temperature, what is the temperature when melting starts?
3. Refer to the Ag–Cu phase diagram as shown in Figure 5.8.
   a. What is the maximum solubility of Cu in silver?
   b. What is the composition of the eutectic alloy?
   c. For a silver–copper alloy with 10 w/o of Cu, what is the temperature at which liquid starts to appear as we increase the temperature gradually from room temperature?
   d. For the above alloy, at thermal equilibrium at 800°C (two-phase region), estimate the weight fraction of the liquid.
4. a. Sketch the phase diagram of a binary system in which the two components are completely miscible through the entire composition range (e.g., Ni and Cu). This phase diagram should indicate the solid phase, the liquid phase, and the two-phase region.
   b. Pick any intermediate composition for the solid phase. Indicate on your phase diagram the temperature at which the solid phase starts to melt.
   c. Pick the same composition but in the liquid region. Indicate on the phase diagram the temperature at which the liquid phase starts to solidify.
5. Refer to the Pb–Sn phase diagram in the text. Start with an alloy containing 80 w/o tin, which is first melted and then allowed to cool slowly to room temperature. Sketch the microstructure in the $\beta + L$-region as well as the final microstructure at room temperature.
6. Refer to the Al–Si phase diagram in Figure 5.22.
   a. Express in weight percent the maximum amount of aluminum that can be dissolved in silicon. At what temperature does this occur?
   b. Starting at 1200°C with a 50:50 alloy and cooling, state the temperature at which solidification begins. Does the solid contain mostly silicon or mostly aluminum? At 800°C, estimate the weight fraction of the solid phase.
   c. What is the eutectic temperature and the eutectic composition in weight percent of Si?
7. Hypo- and hypereutectoid steels are distinguished by the carbon content:
   a. What is the approximate carbon content that divides these steels?
   b. State one desirable property of hypoeutectoid steels. Why do these steels have this property?
   c. State one desirable property of hypereutectoid steels. Why do these steels have this property?
8. A certain town in the Midwest has 60 miles of roadway, with an average width of 50 ft. Assuming that NaCl is to be used for melting the snow, estimate how much salt should be used to handle a 6-in.-snow event. For the purpose of this estimate, you may assume that the density of snow is equal to 10% of liquid water (density of water = 62.4 lb/ft³) and that the ambient temperature is around 20°F. This calculation shows why snowplows are always needed.
9. During the phase transformation process at a given temperature, a fraction of the new phase z appearing as a function of time t can be written as:

$$z = 1 - \exp(-kt^n),$$

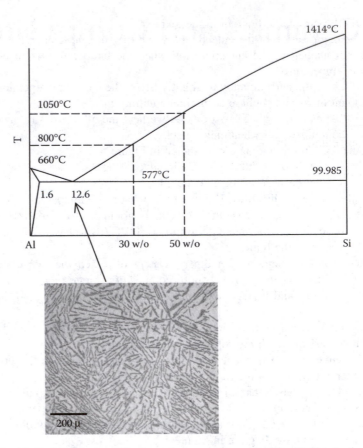

FIGURE 5.22 The Al–Si phase diagram. The inset shows a micrograph of the Al–Si eutectic. Note the lamellae composed of silicon- and aluminum-rich phases. (Micrograph reprinted with permission from Kathleen Stair.)

where k and n are constants. Define t_{10} and t_{90} as the time when 10 and 90% of the new phase appears, respectively. Given that $t_{10} = 11.0$ s and $t_{90} = 23.8$ s, determine n and k. Sketch z as a function of time t.

10. Refer to the Ag–Cu phase diagram in Figure 5.8. For a polycrystalline Ag–Cu alloy obtained by slow cooling from the melt, describe and sketch how the microstructure differs for each of these alloys: 10, 28.1 (eutectic), and 90 w/o Cu.

6 Ceramics and Composites

6.1 RECIPE FOR ICE FRISBEES

Since ceramic materials are often associated with pottery and cookware, let us start our discussion by cooking up something with the following recipe*:

1. Get two small plates.
2. Fill both plates halfway with water.
3. Get a piece of paper (tissue or magazine) and break it into small (5 mm) pieces.
4. To one plate, add these small pieces of paper and mix well.
5. Put both plates into the freezer overnight.
6. Remove the plates from the freezer.
7. Release the two ice Frisbees from the plates.

At this point, you may wish to go outdoors. Throw both Frisbees to a height of 10–15 ft (be careful — do not try to catch this Frisbee). What do you find (Figure 6.1)? What has this to do with ceramics? Why are these findings significant?

Before we get back to these questions, let us define what ceramic materials are. Traditionally, ceramics are materials that come out of clay after high-temperature firing: china, porcelain, tiles, bricks, and glassware (*keramikos* = burnt material). Today, we consider ceramics to be the general class of inorganic nonmetallic compounds (semiconductors excluded). Usually, a ceramic compound consists of at least one metallic element and one nonmetallic element. Such ceramics are partially ionic. As we will see later, this has a large effect on the electrical and mechanical properties of ceramics.

Returning to the ice Frisbee experiment, we can consider water (hydrogen oxide) to be a special case of an oxide ceramic. The Frisbee-throwing experiment should result in the pure oxide ceramic breaking into pieces upon the first throw (Figure 6.1). This is not surprising since we learned in an earlier chapter that ceramics tend to have lower fracture toughness and hence fail more easily with sudden impact. On the other hand, the paper-impregnated oxide ceramic can withstand multiple throws without failure. How is it possible that something as fragile as paper can impart such dramatic performance improvement to this oxide ceramic? This is a classic case of a composite material that retains the strength of the matrix, but with markedly improved fracture toughness. We will discuss this point further in a later section. It is left as an exercise for the reader to determine what happens when the amount of paper is changed or replaced with metal paper clips or little pieces of rubber bands.

6.2 CRYSTAL STRUCTURES

Before we attempt to explain results obtained from the ice Frisbee experiment, let us spend a few moments exploring crystal structures of ceramics. Since a ceramic compound has at least two elements and is usually partially ionic, the atomic arrangements are more complicated than for elemental solids. The atomic packing in ceramics generally obeys two rules: (1) The system must

---

* I learned this recipe from Prof. Lynn Johnson at Northwestern University.

Ice   Ice with shredded paper

FIGURE 6.1 Two ice Frisbees, one with embedded paper strips, before and after being dropped 10 ft.

be electrically neutral; and (2) cations and anions are packed to obtain the maximum coordination. The second rule simply means that each cation would like to bind to as many anions as possible and vice versa. In this way, the system attains the lowest electrostatic energy. The maximum coordination depends on the cation-to-anion radius ratio (r_C/r_A), where r_C is the cation radius and r_A the anion radius.* Table 6.1 shows that there is a unique crystal structure corresponding to a given cation-to-anion radius ratio.

To illustrate this point, let us calculate the r_C/r_A ratio for a coordination number of three (Figure 6.2). At the limit of stability, the central cation is contacting all three anions, and the three anions are barely touching each other, as shown. From the dotted triangle, we can write:

$$\frac{r_A}{r_C + r_A} = \cos 30^o = \frac{\sqrt{3}}{2}.$$

Therefore,

$$\frac{r_C}{r_A} = \left(\frac{r_A}{r_C + r_A}\right)^{-1} - 1$$

$$= \frac{2}{\sqrt{3}} - 1 = 0.155.$$

---

* Generally, the cation radius is less than the anion radius.

TABLE 6.1
Relationship among Cation/Anion Radius Ratio, Coordination
Number, and Geometry

r_C/r_A	Coordination Number	Coordination Geometry
<0.155	2	
0.155–0.225	3	
0.225–0.414	4	
0.414–0.732	6	
0.732–1.0	8	
>1.0	12	

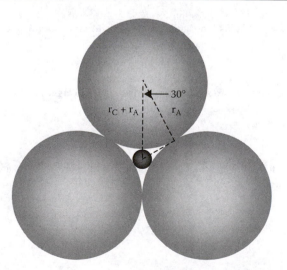

FIGURE 6.2 Stability configuration for a coordination number equal to three. The center circle is the cation (radius = r_C) and the outer three circles are anions (radius = r_A).

TABLE 6.2
Some Common Ceramic Structures

Name	Example	Coordination Number	Structure
Rock salt	NaCl (simple cubic) FCC	6	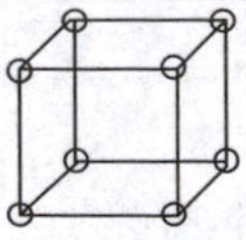
Cesium chloride	CsCl (BCC)	8	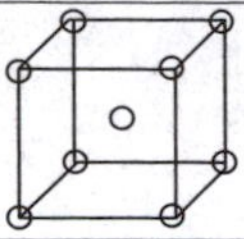
Zinc blende	GaAs (diamond cubic)	4	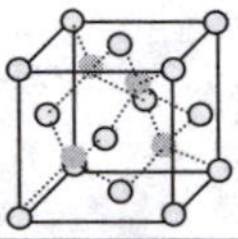
Perovskite	$SrTiO_3$	6 for Ti	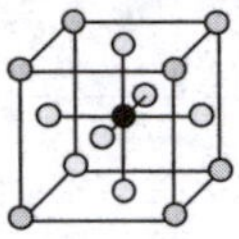

When the cation radius is too small, such that $r_C/r_A < 0.155$, the cation is unable to contact all three anions. Threefold coordination becomes unstable, and the structure collapses to twofold coordination. When $r_C/r_A > 0.155$, cation–anion contact is maintained. The threefold coordination is stable as r_C/r_A increases, until a more stable configuration (fourfold tetrahedral coordination) becomes possible. This occurs when $r_C/r_A > 0.225$. Examples of some common ceramic structures are given in Table 6.2.

EXAMPLE

Calculate the cation-to-anion radius ratio at the limit of stability for coordination number equal to eight.

Solution

Refer to Figure 6.3 and Table 6.1 (coordination number equal to eight). In this case, we can write:

$$2r_A + 2r_C = \sqrt{3}a,$$

where a is the unit cell dimension. At the limit of stability, neighboring anions are barely touching each other as shown. Therefore,

$$2r_A = a,$$

so that:

$$2r_A + 2r_C = 2\sqrt{3}r_A.$$

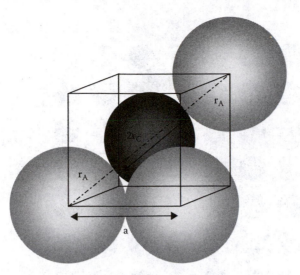

FIGURE 6.3 Stability configuration for a coordination number equal to eight.

Dividing by r_A on both sides of the equation, we have:

$$2 + 2\frac{r_C}{r_A} = 2\sqrt{3}.$$

Solving,

$$\frac{r_C}{r_A} = \frac{2\sqrt{3}-2}{2} = 0.73.$$

6.3 IMPERFECTIONS

6.3.1 POINT DEFECTS

Vacancies and interstitials occur in ceramics and in elemental solids. As discussed in the chapter on imperfections, these defects are produced by thermal excitation or energetic particle bombardment. In ionic ceramics, electrical neutrality must be maintained; such defects can be classified into three types (Figure 6.4):

- Schottky defect: cation or anion moving to the surface, producing cation or anion vacancy respectively
- Frenkel defect: cation or anion moving to an interstitial site, producing cation or anion vacancy respectively
- Antisite defect: a cation occupying an anion site or anion occupying the cation site (usually in covalent or weakly ionic ceramics)

6.3.2 IMPURITIES

Impurities can dissolve in ceramics as solid solutions, similar to metals and other elemental solids. The size of the impurity ion determines the solubility, and its charge determines the resulting defects in the ceramic. For example, the radius of Ti^{4+} (as in TiO_2) is 0.068 nm, and the radius of Al^{3+} (as in Al_2O_3) is 0.050 nm. The difference is more than 30%. As a result, there is no mutual solubility between these two oxides.

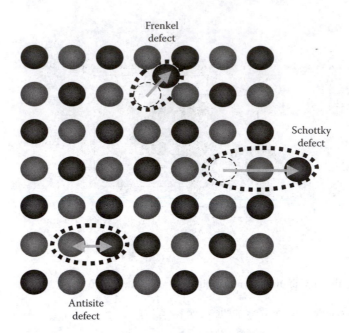

FIGURE 6.4 Point defects in ionic solids.

To illustrate the effect of impurity charge state on defects, consider adding small amounts of $CaCl_2$ to NaCl. For each Ca^{2+} going into a Na^+ site, we produce an extra positive charge. To maintain overall electrical neutrality of the crystal requires the generation of one cation vacancy for each Ca^{2+} introduced. Another way to look at this problem is to consider a sodium chloride crystal with exactly 100 cation (+) sites and 100 anion sites. One $CaCl_2$ molecule is introduced. All 100 anion sites are filled with chloride ions (Cl^-). We need to populate the cation sites so that the total positive charge is exactly +100e. This can be accomplished with 98 Na^+ and 1 Ca^{2+}, for a total of 99 cation sites, leaving 1 cation site vacant.

6.4 MECHANICAL PROPERTIES

6.4.1 Brittle Fracture of Ceramics

Even without doing the ice Frisbee experiment, we know it is a bad idea to drop glassware or ceramic plates onto the floor because of their brittle nature. The primary reason for this brittleness is that dislocations do not move easily in ceramics at or near room temperature. As a result, little or no plastic deformation occurs prior to the stress exceeding the tensile strength. Brittle fracture occurs with little warning. This mode of failure is undesirable for structural applications.

To understand why plastic deformation is difficult in ceramics, refer to Figure 6.5, which shows the displacement of atoms along two directions in a sodium chloride structure. When you attempt to slide one part of the crystal along the [100] direction as shown, you may notice that ions of the same sign face each other after sliding past one half the repeat distance in that direction. This results in a significant energy barrier, so slip along that direction is not favorable. On the other hand, slip along the [110] direction as shown is much easier; positive and negative ions always face each other during slip. In general, fewer slip systems are available in ceramics compared with metals.*

---

* In FCC metals, 12 slip systems are available. In most ceramics, perhaps only 4–6 slip systems are available, depending on the structure.

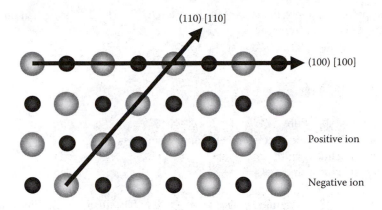

FIGURE 6.5 Slip along (100)[100] and (110)[110] directions in a NaCl structure.

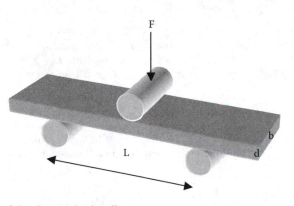

FIGURE 6.6 Geometry of the three-point bending test.

Although slip along the [110] direction is allowed, the magnitude of the slip vector b (Burgers vector) is longer than that in typical metals (e.g., 0.392 nm for sodium chloride versus 0.255 nm for copper). Since dislocation energy is proportional to $|b|^2$, it is more difficult to create dislocations in ceramics compared with metals. These two factors (fewer slip systems available and difficulty of generating the required dislocations) explain why plastic deformation is difficult in ceramics. As stated earlier, without plastic deformation, brittle fracture occurs suddenly and without warning.

For materials exhibiting plastic deformation, the large stress at the corner of a sharp crack or flaw causes dislocation motion, resulting in blunting of the sharp crack or flaw. In the absence of plastic deformation, a crack tip or internal flaw in ceramics does not get blunted under crack-opening stress, causing an excessive stress ahead of the crack tip.* As a result, ceramics are weaker under tension than compression, and their mechanical properties are more sensitive to the existence of flaws. The mode-I fracture toughness K_{Ic} of common ceramics is ≤ 1–5 MPa-m$^{1/2}$, compared with ~30–100 MPa-m$^{1/2}$ for many aluminum alloys and steels.

6.4.2 Flexural Strength

Because of the brittleness of ceramics, it is difficult to measure the mechanical strength by normal tensile testing, which requires mounting such specimens in tight grips without fracturing them.

---

* Glass windows in commercial aircraft are sandwiched between plastic protectors. This is to avoid having accidental scratches on the glass surface (due to abrasion from debris or passengers' sticky fingers). Such scratches can act as stress concentrators that may have disastrous consequences.

Instead, a three-point bending scheme is used (Figure 6.6). The load F is applied until failure. The flexural strength σ_f is given by:

$$\sigma_f = \frac{3F_f L}{2bd^2},$$ (6.1)

where

F_f = the load at failure
L = the length of the specimen between the supports
b = the specimen width
d = the specimen thickness

Flexural strength of typical ceramics ranges from ~100 MPa (silica, magnesia) to over 500 MPa (alumina, silicon carbide, zirconia, and silicon nitride).

EXAMPLE

A piece of silicon carbide is loaded as shown in Figure 6.6. The dimensions are as follows: width b = 1 cm; specimen thickness d = 0.2 cm; distance between supports L = 5 cm. The sample fails at an applied load of 100 N. Determine the flexural strength of the silicon carbide sample.

Solution

The first step is to make sure that we keep consistent units. In this example, b = 0.01 m, d = 0.002 m, L = 0.05 m, and F_f = 100 N. Substituting these numbers into Equation (6.1), we calculate the flexural strength σ_f to be:

$$\sigma_f = \frac{3F_f L}{2bd^2} = \frac{3 \times 100 \times 0.05}{2 \times 0.01 \times (0.002)^2} \, \text{Pa}$$

$$= 1.9 \times 10^8 \, \text{Pa}$$

$$= 190 \, \text{MPa}.$$

6.4.3 THERMAL SHOCK RESISTANCE

Have you ever experienced the cracking of your favorite ceramic mug when pouring hot water into it? When the surface of the ceramic is exposed to sudden temperature increase, it expands relative to the bulk, which takes time to attain the same temperature due to the low thermal conductivity of most ceramic materials. This expansion is equivalent to the application of a tensile stress to the ceramic material and is known as thermal stress, given by:

$$\sigma_{th} = \frac{E\alpha\Delta T}{1-\nu},$$ (6.2)

where

σ_{th} = the thermal stress
E = Young's modulus
α = coefficient of thermal expansion
ΔT = temperature change
ν = Poisson ratio

Equation (6.2) is applicable when the thermal stress is biaxial — that is, acting parallel to the surface of the material. For ceramic materials subjected to sudden heating or cooling, failure comes from tensile thermal stress acting on surface flaws, which act as stress concentrators.

EXAMPLE

Consider alumina (Al_2O_3). Its coefficient of thermal expansion is 8.8×10^{-6}/K. The Young's modulus is 380 GPa and the Poisson ratio is 0.3. Calculate the thermal stress for a temperature change of 250 K.

Solution

Substituting the appropriate values into Equation (6.2), we have:

$$\sigma_{th} = \frac{380 \times 10^9 \times 8.8 \times 10^{-6} \times 250}{1 - 0.3}$$

$$= 1.2 \times 10^9.$$

Therefore, the thermal stress is equal to 1.2 GPa, substantially greater than the flexural strength (350–1000 MPa) of alumina.

A material with high thermal conductivity minimizes problems with thermal stress since heat conduction helps to equalize surface and bulk temperatures of the sample. Therefore, the temperature difference term (ΔT) in Equation (6.2) can be assumed to be inversely proportional to the thermal conductivity so that:

$$\sigma_{th} = \frac{E\alpha\Delta T}{1 - v} \propto \frac{E\alpha}{\kappa}, \tag{6.3}$$

where κ is the thermal conductivity. This shows that in order to obtain lower thermal stress, we must maximize the term $\kappa/E\alpha$. In the ceramics community, this term is known as the thermal shock resistance, a measure of the ability of the material to withstand sudden heat input without fracture. Larger $\kappa/E\alpha$ values imply better thermal shock resistance.

EXAMPLE

Compare the thermal shock resistance of two ceramic materials:

Alumina: $\kappa = 40$ W/m-K, $E = 380$ GPa, $\alpha = 8.8 \times 10^{-6}$/K
Fused silica: $\kappa = 1.4$ W/m-K, $E = 70$ GPa, $\alpha = 0.5 \times 10^{-6}$/K

Solution

For alumina, the thermal shock resistance (TSR) is given by:

$$TSR = \frac{40}{380 \times 8.8 \times 10^{-6}} \approx 1.2 \times 10^4.$$

For fused silica, the corresponding thermal shock resistance is given by:

$$TSR = \frac{1.4}{70 \times 0.5 \times 10^{-6}} \approx 4 \times 10^{4}.$$

Therefore, fused silica has better thermal shock resistance than alumina. This is primarily due to the lower thermal expansion coefficient of fused silica. As inferred from Equation (6.3), ceramic materials with low thermal expansion coefficient and high thermal conductivity have high thermal shock resistance.

Question for discussion. Sometimes, it is healthy to challenge accepted practice. In the preceding discussion, the parameter $\kappa/E\alpha$ is considered to be a measure of the thermal shock resistance. However, when the ceramic material is sufficiently thin, the temperature difference between the surface and the sample interior becomes negligible. The thermally induced surface stress is therefore minimal. The parameter $\kappa/E\alpha$ does not take this into account. Is there a better figure of merit for thermal shock resistance?

6.4.4 INFLUENCE OF POROSITY

Most ceramic materials are made by compacting the ceramic materials in powder form and then heating them to elevated temperatures to cause fusion of these powders. This fusion process is known as sintering. Some porosity may remain after sintering. The resulting porosity acts as flaws, which produce stress concentration and result in failure at lower applied stress according to:

$$\sigma_f = \sigma_o \exp(-nP), \tag{6.4}$$

where

σ_f = the flexural strength
σ_o = the strength of the ceramic material without porosity
P = the volume fraction of porosity
n = an empirical constant

Because porosity is empty space, the elastic modulus is also reduced. The following empirical relationship between elastic modulus and volume fraction of porosity is found:

$$E = E_o(1 - 1.9P + 0.9P^2), \tag{6.5}$$

where E is the elastic modulus, E_o the elastic modulus of a porosity-free solid, and P the volume fraction of porosity.

One way to minimize porosity is the development of glass ceramics.* As the name implies, the starting point is glass, which can be considered to be a supercooled liquid with an amorphous structure. Most glasses are made of silica–alumina. By introducing an appropriate modifier and with careful heat treatment, glass ceramics can be produced with 90% of the materials crystallized and having grain sizes of ~0.1–1 μm. The remaining 10% glass phase fills the porosity between crystal grains. Without such porosity, glass ceramics tend to be stronger and have higher thermal conductivity; they are therefore less susceptible to thermal shock.

---

* One of the most well-known glass ceramics is Corning cookware. The primary components are SiO_2, Al_2O_3, and Li_2O (modifier). With much improved thermal shock resistance, this type of cookware has been advertised to go from the refrigerator to the oven without harm.

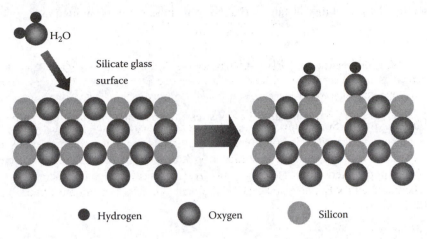

FIGURE 6.7 Schematic illustration of the attack of the silicate network by water.

6.4.5 ENVIRONMENTAL EFFECTS

The environment can have dramatic effects on the mechanical properties of ceramics, similar to metals. For example, it is much easier to fracture glass after scribing and then wetting the scribe mark, a fact known to glassblowers and cutters. Most silica glasses have the basic SiO_4^{4-} units, in which silicon sits at the center of a tetrahedron and is surrounded by four oxygen ions. Water reacts with the Si–O–Si network to form Si–OH, effectively opening up the network (Figure 6.7). Under stress, this facilitates the nucleation of cracks and their propagation to failure.

Another example is sodium chloride. It is a strong ionic solid, as indicated by its high melting point (801°C) and large cohesive energy (786 kJ/mol). As is common knowledge, in spite of such strong bonding, it can be destroyed readily by water.

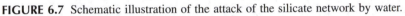

Statistics Note

Mechanical properties of ceramics have more statistical scatter than metals due to the presence of and their sensitivity to flaws. One way to characterize this statistical scatter is to construct a Weibull plot — that is, log(percentage of samples failed at a given stress) versus log(stress), as shown in Figure 6.8. The slope of the fitted straight line is known as the Weibull modulus. For a typical ceramic material, the Weibull modulus is smaller than that for a typical metal.

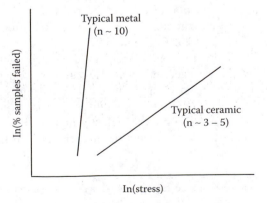

FIGURE 6.8 Schematic diagram of the Weibull plot.

Generally, the probability of failure P at a given stress σ can be written as:

$$P(\sigma) = 1 - \exp\left[-\left(\frac{\sigma}{\sigma_o}\right)^n\right], \tag{6.6a}$$

where σ_o and n are material constants. At low stress, the exponential function can be approximated as:

$$P(\sigma) \approx \left(\frac{\sigma}{\sigma_o}\right)^n. \tag{6.6b}$$

Therefore, we can write:

$$\ln P(\sigma) \approx n \ln \sigma + C, \tag{6.7}$$

where C is some constant.

Therefore, for low stress, the slope of $\ln P$ versus $\ln \sigma$ gives directly n, the Weibull modulus. Larger n implies smaller statistical scatter, which is desirable. The Weibull plot provides a useful design tool to choose an appropriate stress level for a given application.

6.5 TOUGHENING OF CERAMICS

Ceramics have many useful attributes: high temperature stability, chemical inertness, hardness, etc. However, one major disadvantage is the brittleness of ceramics at ambient temperatures. Many research investigations have explored different strategies to toughen ceramics. We will discuss four such strategies here.

6.5.1 TRANSFORMATION TOUGHENING

A good example of transformation toughening is the stabilization of zirconia by the addition of CaO. At room temperature, pure ZrO_2 has the monoclinic structure, changing to tetragonal above 1150°C and to cubic above 2350°C. When cooled to below 1150°C, pure zirconia undergoes the tetragonal-to-monoclinic transformation. A volume expansion of about 4% occurs during this transformation, resulting in the formation of cracks.* To prevent this, 3–7 w/o CaO is typically added to zirconia. According to the zirconia–calcia phase diagram, this results in a mixture of tetragonal and cubic phases above ~1000°C (Figure 6.9).

When cooled from, say, 1000°C to room temperature at normal rates, the tetragonal zirconia particles are constrained by the stronger matrix (cubic zirconia) and do not transform to the monoclinic phase. Therefore, there is no abrupt volume change, and no cracks are formed. This is known as partially stabilized zirconia (PSZ). Other additives such as yttrium oxide and magnesium oxide have also been used for stabilization (at sufficiently high additive concentrations, only the cubic phase is retained at room temperature, and the material is then known as fully stabilized zirconia).

---

* This is one of several instances in which a solid expands with decreasing temperature. The transformation of FCC to BCC iron is another.

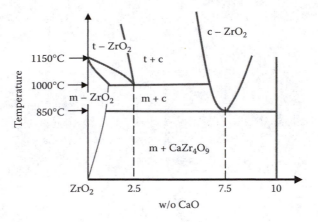

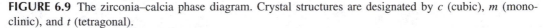

FIGURE 6.9 The zirconia–calcia phase diagram. Crystal structures are designated by *c* (cubic), *m* (monoclinic), and *t* (tetragonal).

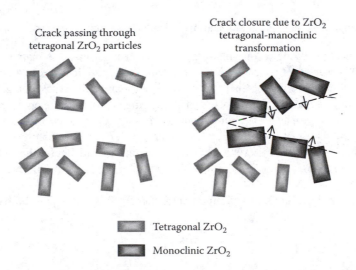

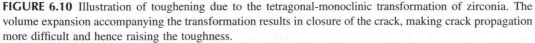

FIGURE 6.10 Illustration of toughening due to the tetragonal-monoclinic transformation of zirconia. The volume expansion accompanying the transformation results in closure of the crack, making crack propagation more difficult and hence raising the toughness.

Now consider a crack or flaw inside the PSZ or a ceramic matrix (e.g., alumina or zirconia) with embedded PSZ particles. As the crack passes through the tetragonal zirconia particles, the particles become unconstrained and immediately transform to the monoclinic phase. The transformation results in a large (4%) volume expansion, which creates a local compressive stress, effectively closing the crack. This makes crack propagation more difficult and raises the fracture toughness to significantly higher levels ($K_{Ic} \sim 6\text{--}10$ MPa-m$^{1/2}$) (Figure 6.10).

6.5.2 FIBER OR PARTICULATE REINFORCEMENT

The ice Frisbee experiment shows that the addition of paper increases the fracture toughness of ice. This is the second method of toughening. Here, a second phase is introduced into the ceramic matrix in the form of fibers or particulates (Figure 6.11). The second-phase material increases the fracture toughness through three mechanisms:

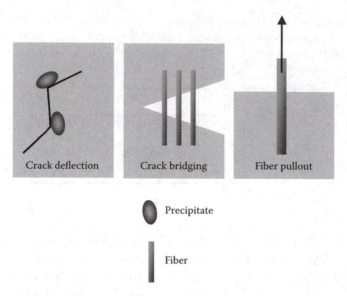

FIGURE 6.11 Toughness-enhancement mechanisms in particle- and fiber-reinforced ceramics.

- Crack deflection. If the second-phase material is stronger than the matrix, it will deflect cracks. With sufficient concentration of second-phase particles, cracks zigzag within the material, thus increasing the energy required to propagate the crack in a given direction.
- Crack bridging. Second-phase fibers can form bridges across the opening crack. In order for the crack to propagate, additional energy must be dissipated to break the fibers.
- Fiber pullout. During tensile deformation, fibers may be pulled out of the matrix. Depending on the bonding characteristics between the matrix and fibers, the pull-out process requires overcoming the friction or breaking bonds at the fiber–matrix interface.

All three processes require additional dissipation of energy, thus increasing the fracture toughness of the ceramic.

6.5.3 CERMETS

A cermet is simply a mixture of ceramics and metals, a special case of composite materials. The intent is to combine the high strength of ceramics with the high fracture toughness of metals. A good example is WC–Co. Adding 5–10% cobalt (a tough metal) to tungsten carbide (a strong and wear-resistant ceramic) results in a material with excellent wear resistance under high stress. This material is known as cemented tungsten carbide and is used in cutting tools.

6.5.4 SURFACE MODIFICATION

As discussed earlier, ceramics are stronger under compression than under tension. Therefore, another way to toughen ceramics is to put the surface under compression by some means. If any tensile stress is applied to the ceramic, it has to overcome the imposed compressive stress first before the surface is placed under tension. This would minimize the potential of surface flaws to form fatal cracks.

Two techniques are commonly used to impose the surface compressive stress. One is to introduce impurities to the surface with atomic size larger than the matrix — for example, introducing potassium ions to replace sodium ions on the surface of glass or chromium ions to replace aluminum ions on the surface of sapphire (Al_2O_3). Another technique is to adjust the cooling rate from the melt. This technique is commonly applied to glass. When the surface is suddenly cooled from the melt, it turns solid while the inside of the glass is still liquid. Subsequent slow cooling causes the inside to become solid as well. The liquid–solid transformation results in a volume contraction, which in turn results in a compressive stress on the glass surface. This type of glass is known as tempered glass and is widely used in cars and prescription lenses.

6.6 ELECTRICAL, MAGNETIC, OPTICAL, AND THERMAL APPLICATIONS

Ceramic materials have a diverse range of properties and are used in many different applications. The following sections give a sampling of such applications:

6.6.1 ELECTRICAL INSULATORS

Many oxides such as silicon oxide and aluminum oxide are excellent insulators and can withstand large electric fields without breakdown. Common ceramic materials such as fused silica, Pyrex glass, and mica have breakdown fields (also called dielectric strengths) in the range of $1-4 \times 10^7$ V/m. The basis of the silicon integrated circuit technology relies on the easy formation of thin insulating silicon dioxide layers with high dielectric strength.* For example, high-quality silicon dioxide thin films can achieve dielectric strength in excess of 5×10^8 V/m. Therefore, a 10-nm thick SiO_2 film can sustain the application of a voltage of 5 V or less without breakdown.

Dielectric breakdown poses another limit to how small integrated circuits can be. Thermal energy at room temperature ($3/2k_BT$) provides an equivalent voltage of 40 mV. At this voltage, breakdown occurs at a silicon oxide thickness of about 0.08 nm, assuming a dielectric strength of 5×10^8 V/m.

Insulating ceramics can be thought of as large bandgap semiconductors. In the absence of impurities or defects, the conduction electron concentration is proportional to $\exp(-E_{gap}/2k_BT)$, where E_{gap} is the bandgap energy, k_B the Boltzmann constant, and T the absolute temperature. Therefore, a large bandgap means small conduction electron concentration and hence low electrical conductivity. When an external electric field is applied, conduction electrons will be accelerated. The energy gain is given by:

$$\Delta E = \frac{1}{2}\frac{e^2 E^2 \tau^2}{m}, \qquad (6.8)$$

where

ΔE = the energy gain
e = the electron charge
E = the electric field
τ = average time between collisions
m = electron mass

When this energy gain is equal to or greater than the bandgap energy E_{gap} and this electron collides with a valence electron, the energy transfer is sufficient to excite the valence electron across the gap to produce a conduction electron. As a result, a collision cascade occurs, leading to a rapid

---

* Not all ceramics are insulators. Some ceramic materials are excellent electrical conductors, such as many transition metal borides, carbides, and nitrides (TiN is used in integrated circuits as a diffusion barrier), boron carbide, ruthenium oxide, silver oxide, etc. From the late 1980s to 1990s, a series of complex cuprates were found to exhibit superconductivity at relatively high temperatures (up to 150 K). In 2001, magnesium diboride was discovered to be a superconductor with a superconducting transition temperature of 39 K.

increase in the electrical current and subsequent breakdown. Therefore, the dielectric breakdown condition is given by:

$$E_{gap} = \Delta E \left(= \frac{1}{2} \frac{e^2 E^2 \tau^2}{m} \right).$$

(6.9a)

The breakdown field E_{bd} is then obtained by solving for E in the preceding equation:

$$E_{bd} = \frac{1}{e\tau} \sqrt{2mE_{gap}} \; .$$

(6.9b)

This suggests that large bandgap materials with short collision times give the highest dielectric breakdown strengths.

6.6.2 CAPACITORS

When a material with dielectric constant ε is introduced into a parallel plate capacitor, the capacitance increases by a factor ε. Certain ceramic materials, such as barium titanate, have very large dielectric constants (>1000). This means that it is possible to obtain large capacitance on small footprints by using materials with large dielectric constants.

Why do some ceramic materials possess high dielectric constants? Barium titanate is a ferroelectric material consisting of permanent electrostatic dipoles (analogous to magnetic dipoles in a magnetic material). In the absence of an external field, these permanent dipoles organize into domains of various orientations so as to minimize the total electrostatic energy. When an electric field E is applied, domains with the favorable orientation grow at the expense of those not in the favorable orientation, thus increasing the net dipole moment per unit volume (known as polarization P). The aligned dipoles increase the overall electric field. The ratio of this overall electric field within the material to the original applied field is the dielectric constant. It can be shown from standard electrostatics that the dielectric constant ε is given by:

$$\varepsilon = 1 + \frac{P}{\varepsilon_o E},$$

(6.10)

where ε_o is the permittivity of free space (8.8×10^{-12} F/m), E the electric field, and P the polarization. The second term, $P/\varepsilon_o E$, is known as polarizability α. Therefore, a material with a large dielectric constant means that its polarization is readily increased by an external electric field.

EXAMPLE

Consider applying an electric field of 1×10^6 V/m to a ceramic material. The resulting polarization is 0.01 Coul/m². Calculate the dielectric constant.

Solution

As given in Equation (6.6), the dielectric constant ε is given by:

$$\varepsilon = 1 + \frac{0.01}{8.8 \times 10^{-12} \times 10^6} \cong 1100.$$

6.6.3 OXYGEN ION CONDUCTORS

Some oxide ceramics become good oxygen ion conductors at moderate temperatures. For example, at 500°C or greater, oxygen anions can migrate inside zirconia at reasonable rates by hopping through oxygen vacancies. Zirconia, as a result, is used as a solid electrolyte in automotive oxygen sensors and hydrogen fuel cells. However, not every oxide works well as an ionic conductor. At elevated temperatures, electron excitation across the bandgap of the oxide provides a source of conduction electrons, which reduces the electrical resistance of the solid electrolyte and short circuits the applied voltage, if present, in sufficiently high concentrations. Therefore, a good oxygen ion conductor should have a large bandgap and low electron mobility.

6.6.4 DATA STORAGE

Some iron-based oxides are ferromagnetic and can be used for magnetic data storage. The earliest generation of computer hard disks (known as brown disks) used γ-Fe_2O_3 as the storage medium. Iron-chromium oxides continue to be used in magnetic tapes. Certain oxides (e.g., barium titanate) are ferroelectric (i.e., having permanent electrostatic dipole moments) and can be used for non-volatile data storage.

6.6.5 OPTICAL FIBERS

Instead of moving data using metallic conductors, we now routinely do so by optical means — for example, in computer networks, telecommunications, and CD/DVD drives. In this case, light is transmitted over optical fibers typically made of high-purity silicon oxide. For telecommunications, in order to minimize the use of repeaters over long distances, the wavelength of light used is such that the absorption coefficient is minimum in silicon oxide, which is ~1.3 µm. The absorption A is measured in terms of decibels (dB) per kilometer:

$$A(dB/km) = 10 \times \log_{10}\left(\frac{I_t}{I_o}\right), \tag{6.11}$$

where I_t is the light intensity after traveling through 1 km of the fiber and I_o the incident light intensity. Therefore, an attenuation of –3 dB/km means that the light intensity is reduced by 50% (–3 ≈ 10 $\log_{10}$0.5) after traveling through 1 km of the optical fiber.

6.6.6 THERMAL INSULATORS

Ceramic materials have a wide range of thermal conductivity. At room temperature, pure copper has a thermal conductivity κ of about 400 W/m-K); the κ value is about 16 W/m-K for 304 stainless steels and about 200 W/m-K for 2024 Al alloys. In comparison, the κ value is about 40 W/m-K for crystalline alumina (sapphire), decreasing to 1.4 W/m-K for fused silica and Pyrex.* Ceramic thermal insulators are used in applications ranging from household appliances to turbine blades and improved insulation for houses.

EXAMPLE

To appreciate what it means for a material to have a thermal conductivity of 1.4 W/m-K, consider the following equation relating heat flux to thermal conductivity:

---

* Thermal conductivity of a given material depends on temperature. For many ceramics at room temperature, the thermal conductivity decreases with increasing temperature.

$$Q = -\kappa \frac{dT}{dx}, \tag{6.12}$$

where Q is the heat flux in J/m²-s, and dT/dx the temperature gradient. The negative sign means that the heat flux is in the opposite direction of the temperature gradient. Consider delivering a heat flux of 1.4×10^6 W/m² through this material (and removing the same amount on the other end). This is equivalent to the energy flux from approximately 1000 suns. Determine the temperature difference across a thickness of 1 mm of this material.

Solution

Substituting with $Q = 1.4 \times 10^6$ W/m², $\kappa = 1.4$ W/m-K, and $\Delta x = 10^{-3}$ m, we have:

$$\left| \frac{dT}{dx} \right| = \frac{Q}{\kappa} = \frac{1.4 \times 10^6}{1.4}$$

$$= 1 \times 10^6 \text{ K/m}.$$

Therefore, the temperature difference ΔT sustained across 1 mm is equal to:

$$\Delta T = 1 \times 10^6 \Delta x$$

$$= 1 \times 10^6 \times 10^{-3}$$

$$= 1 \times 10^3 \text{ K}.$$

This means that we can maintain a temperature difference of 1000 K across a thickness of only 1 mm.*

6.6.7 SMART MATERIALS**

Ceramics such as barium titanate and lead zirconium titanate (PZT) exhibit the piezoelectric effect — that is, strain induced by the application of an electric field. This is due to the distortion of the unit cell so that the center of gravity for the positive charges does not coincide with that for the negative charges. As a result, below a certain critical temperature known as the Curie temperature, a permanent electric dipole exists, as shown in Figure 6.12. Above the Curie temperature, thermal excitation is sufficient to cause the electric dipole moment to vanish.

By applying an electric field along the direction of the dipole, we can change the distortion of the unit cell. Therefore, an electric field can produce mechanical strain. Conversely, an external stress can distort the unit cell and hence affect the electric field (i.e., the piezoelectric effect works in both directions). These materials are used as actuators producing motion with applied voltage (e.g., in precision motion control of advanced microscopes and other scientific instruments) and transducers producing voltage output in response to stress (e.g., piezoelectric microphones, submarine detection, and pressure sensors).

---

* Ceramic coatings, about 100 μm thick, are applied as thermal barriers in turbine blades for use in turbojet aircraft. These coatings provide another 50–100 K increase in operating temperatures to obtain higher energy efficiency.
** "Smart materials" is probably a misnomer. The smartness is derived from control algorithms and systems, based on the response of the material due to an external input.

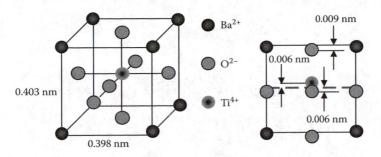

FIGURE 6.12 The barium titanate (BaTiO$_3$) unit cell, showing upward displacement of Ti^{4+} and downward displacement of oxygen ions.

How Big Is It?

The intent of this section is to estimate the size of typical piezoelectric strains. Consider an electrostatic dipole made of $+e$ and $-e$, separated by an equilibrium distance of 0.1 nm. Assume that in addition to Coulomb forces, a short-range repulsive force acts upon these two charges with the form A/R^n, where A is some constant and $n = 10$. An electric field E of 1×10^7 V/m (equivalent to 10 V/μm) is applied to this dipole. The polarity is such that the dipole is stretched. We will calculate the new separation between the $+e$ and $-e$ charges as a result of this external electric field.

Before the application of the external electric field, the following equilibrium condition holds:

$$\frac{e^2}{4\pi\varepsilon_o R_o^2} = \frac{A}{R_o^n},$$

where ε_o is the permittivity of free space (8.8×10^{-12} F/m), and R_o is the equilibrium separation for the dipole.

Solving yields:

$$R_o = \left(\frac{4\pi\varepsilon_o A}{e^2} \right)^{\frac{1}{n-2}}.$$

In the presence of the external field, we have:

$$\frac{e^2}{4\pi\varepsilon_o R^2} = \frac{A}{R^n} + eE,$$

where R is the new separation distance.

Assuming that eE is small compared with other terms, the method of Taylor expansion can be used to solve for eE:

$$eE = \frac{e^2}{4\pi\varepsilon_o R^2} - \frac{A}{R^n}$$

$$\cong \left(\frac{e^2}{4\pi\varepsilon_o R_o^2} - \frac{A}{R_o^n} \right) + \Delta R \left(-\frac{2e^2}{4\pi\varepsilon_o R_o^3} + \frac{nA}{R_o^{n+1}} \right)$$

$$= \frac{\Delta R}{R_o}(n-2)\frac{e^2}{4\pi\varepsilon_o R_o^2},$$

where ΔR is the change in dipole separation.

Direct substitution gives $\Delta R/R_o = 8.8 \times 10^{-6}$. Therefore, the piezoelectric effect affects the interatomic spacing at the parts-per-million level.

6.7 MECHANICAL PROPERTIES OF COMPOSITES

Most engineering materials for structural applications are not made of single-component materials. Rather, they are composites — that is, combinations of two or more distinct components. The motivation for developing composites is to obtain properties not achievable by individual components. For structural applications, high strength without losing fracture toughness is desirable. The ice Frisbee, a composite made of ice and paper, is a good example.

Many aluminum alloys are composites (e.g., incorporation of silicon carbide or titanium diboride into an aluminum matrix). These alloys belong to a broad class of materials known as *metal–matrix composites*. Other composites include *ceramic–matrix* and *polymer–matrix composites*.

We expect the elastic response of a composite material to be some sort of an average between individual components. Consider a composite with two components, the matrix and a second phase. A stress σ is applied to the composite. There are two possibilities: The two components experience the same stress (isostress condition), or they experience the same strain (isostrain condition). For example, for oriented fibers embedded in a matrix, the isostress condition is obtained by applying the stress perpendicular to the longitudinal axis of the fibers, and the isostrain condition is obtained by applying the stress parallel to the longitudinal axis of the fibers. In both cases, we assume that the fiber–matrix interface is intact. The resulting elastic modulus of the composite is given by:

$$\frac{1}{E_c} = \frac{V_m}{E_m} + \frac{V_p}{E_p} \quad \text{for the isostress condition} \qquad (6.13a)$$

and

$$E_c = V_m E_m + V_p E_p \quad \text{for the isostrain condition}, \qquad (6.13b)$$

where E and V represent the elastic modulus and volume fraction, respectively; subscripts c, m, and p represent the composite, matrix, and precipitate, respectively.

EXAMPLE

Derive Equation (6.13a), the elastic modulus of the composite under the isostress condition.

Solution

The isostress condition implies that:

$$\sigma_c = \sigma_m = \sigma_p,$$

where σ is the applied stress.

The total strain experienced by the composite must be shared between the matrix and the precipitate. The percentage shared by each component depends on the projected area acted upon by the applied stress. Assuming that the projected area is proportional to the volume fraction, we have:

$$\varepsilon_c = V_m \varepsilon_m + V_p \varepsilon_p,$$

where ε is the strain. We can then express ε in terms of σ and E as follows:

$$\frac{\sigma_c}{E_c} = V_m \frac{\sigma_m}{E_m} + V_p \frac{\sigma_p}{E_p}.$$

By the isostress condition, all the σ are equal so that they all cancel out, leaving:

$$\frac{1}{E_c} = \frac{V_m}{E_m} + \frac{V_p}{E_p}.$$

Starting from Equation (6.13), we can show mathematically that the elastic modulus of the composite is always smaller than that of the stiffer component. This does not sound very exciting, until we start to explore systems in which the second-phase precipitates or fibers have dimensions below 100 nm. By synthesizing composites with nanoscale precipitates or fibers at appropriate concentrations, we can produce materials with enhanced strength and toughness.

Experiments using thin films show that when both components have nanometer dimensions, the hardness can be enhanced well over the average value. For example, thin films made of alternating nanolayers of titanium diboride and titanium carbide give hardness values of 60 GPa (Figure 6.13), significantly higher than those of titanium diboride (33 GPa) and titanium carbide (25 GPa). As a comparison, the hardness of most diamond thin films ranges between 50 and 80 GPa. The reason for these enhanced mechanical properties is the presence of interfaces, leading to strengthening mechanisms not considered under the isostress or isostrain condition discussed earlier.

6.8 BIOMEDICAL APPLICATIONS

In spite of the brittleness of ceramics, we are beginning to see the use of ceramics in prosthetic implant applications. Processing of ceramics such as silicon nitride is sufficiently advanced that internal flaw size and porosity become negligible. Therefore, there is less concern about stress amplification by such internal flaws and hence catastrophic failure. Ceramic materials have several advantages over metallic alloys as implants: They have better resistance against wear and corrosion, and the density of ceramics is more closely matched to human bones than that of metals.

Graphite/silicon carbide composites are used in artificial heart valves. The surface of the composite is silicon carbide, which imparts superior wear resistance to the artificial heart valve. The interior is made of graphite, which improves the fracture toughness.

Because of the concern about mercury poisoning as well as the metallic appearance, there is a trend to move away from mercury amalgams in dental restoration. Ceramic powders mixed with polymer binders (more about polymers in another chapter) appear to be the logical replacement.

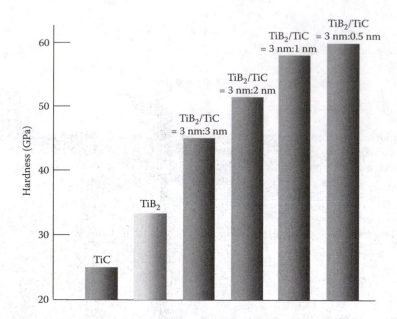

FIGURE 6.13 Hardness values for various TiB_2, TiC, and TiB_2/TiC composite thin films. The notation "TiB_2/TiC = x nm: y nm" means that the film is made of alternating layers of TiB_2 and TiC, of thickness x and y nm, respectively.

In this case, the ceramic grains are analogous to Ag_3Sn and the polymer binder to mercury in amalgams (see the chapter on phase diagrams). Ceramic composites are more resistant to wear and corrosion. The color of ceramic composites is almost identical to natural tooth enamel and therefore aesthetically more appealing than mercury amalgams. As a result, such ceramic composites are becoming the materials of choice among dentists and patients.

PROBLEMS

1. Mechanical properties of ceramics are more sensitive to imperfections such as surface cracks and internal flaws than metals and alloys are. As a result, ceramics have lower fracture toughness.
   a. Explain why ceramics are so sensitive to imperfections such as surface cracks and internal flaws.
   b. In addition to minimizing surface cracks and internal flaws, describe any two methods to improve fracture toughness of ceramics.
2. Some ceramics have strong ionic character. Positive ions like to be closely packed next to negative ions. With this criterion, in order to have a coordination number of six (common in ceramics) in a binary compound made of cations and anions, prove that the ratio of the cation radius to the anion radius must be at least 0.414.
3. MnO and MnS have the NaCl structure. Do you expect the solubility of MnS in MnO to be high or low? Explain.
4. When mechanical properties (e.g., flexural strength) of a given ceramic material are measured, the results have more statistical scatters than those obtained from metallic alloys. Explain.
5. Ceramic dinnerware is often coated with a glaze, which has a slightly smaller coefficient of thermal expansion than the porcelain. The glaze not only improves the appearance, but also makes the dinnerware stronger and resistant to crack propagation. Explain.

FIGURE 6.14 B747 executing the famous "checkerboard" approach to Hong Kong Kai Tak Airport, making a 47° right turn to line up with runway 13 in the final seconds of the approach.

6. Typical commercial glass (e.g., used in windows) contains about 15% sodium. Through an ion-exchange process, sodium ions (ionic radius = 0.098 nm) on the surface can be replaced with potassium ions (ionic radius = 0.133 nm). The strength of the glass improves considerably after this process. Explain.

7. A given ceramic component is to be used in an application in which the maximum tensile stress is σ_{max}. Discuss the fracture toughness, minimum flaw size, and porosity requirements in using a given ceramic material for this application.

8. Figure 6.14 shows a Boeing 747 performing the famous "checkerboard" approach to runway 13 of the Hong Kong Kai Tak International Airport (which was closed in 1997). This runway extends into the harbor (see Figure 1.1 in Chapter 1). Therefore, it is imperative that the aircraft be stopped within a reasonable distance (planes have been known to slide off the runway into the harbor). Consider an aircraft weighing 225,000 kg with a touchdown speed of 65 m/s. Assume that all the kinetic energy is converted to heat in the brakes.

   a. Calculate the total heat energy going into the brakes.

   b. Assuming that the mass of the brakes is 1000 kg and that the specific heat of the brake material is 700 J/kg-K, estimate the increase in brake temperature.
   (Hint: The temperature increase is equal to the total heat energy input divided by the heat capacity of the material [mass × specific heat].)

   c. Discuss the criteria in choosing an appropriate brake material for this application. (Hint: You want the brakes to last, right?)

   d. Considering the following heat capacity data (J/kg-K) for different materials, determine which material should be used for this application.

Al	920
W	133
Al_2O_3	160
Graphite	710

9. The following lists the radii for several divalent metal ions. Given that the radius of O^{2-} is 0.14 nm, deduce the likely crystal structure of each of these divalent metal oxides:

   Mn: 0.067 nm
   Ni: 0.069 nm
   Co: 0.072 nm
   Fe: 0.077 nm
   Ba: 0.136 nm

10. Many prosthetic implant materials (e.g., in hip- and knee-joint applications) use metals and alloys. One reason for using metals and alloys is their high fracture toughness. On the other hand, ceramic materials have lower fracture toughness, but tend to be more inert against corrosion and have better wear resistance under normal dry sliding conditions. A researcher decided to measure the wear performance of zirconia by performing sliding experiments under saline solution (0.09% NaCl) to simulate body fluid conditions. To her surprise, she discovered that zirconia wears rapidly in saline solution. In fact, the wear rate does not appear to depend strongly on the sodium chloride concentration. Discuss the likely wear mechanism of zirconia in this case.*

11. Coatings are being used as thermal barriers to protect turbine blades operating at 1000 K. The heat flux load is 1000 W/cm². To maintain proper tolerance for the physical dimensions of the turbine blade, coatings thicker than 100 μm cannot be used. The requirement is to provide a minimum temperature difference of 50 K across the thermal barrier coating. What coating material(s) will you select?

Material	Thermal Conductivity (W/m-K) at 1000 K
A	5
B	10
C	22

12. Figure 6.7 shows the unzipping of the Si–O–Si network by water. In order for this to occur, it is necessary to break the O–H bond in water and at least one O–Si bond in the silicate network. Where does this energy come from? What is the overall energy balance for the unzipping reaction?

13. Derive Equation (6.13b), the elastic modulus of a composite under the isostrain condition. In this problem, you may assume that the projected area acted upon by the stress on each component is proportional to the volume fraction.

14. a. For composite materials under the isostress condition in which $E_m < E_p$, show that $E_c < E_m$.

    b. For composite materials under the isostrain condition, show that E_c lies between E_m and E_p.

---

* There is an epilogue to this story. Body fluid is more than just water. It contains proteins that interact with surfaces and affects wear of the ceramic materials. In fact, using bovine serum to simulate body fluids, researchers show that zirconia wears much more slowly. Ceramic implant materials are gaining acceptance in Europe.

7 Polymers

7.1 RUBBER BAND EXPERIMENTS

Before starting our discussions on polymers, let us do some simple experiments with rubber bands. Get one of those wide rubber bands — those used by the post office to tie mail should work fine. Stretch the rubber band about two to three times its natural length and hold it there. Get a rough idea of its temperature by touching the rubber band with your lower lip. You may notice that it gets slightly warm. Surprised? Now, relax the rubber band quickly to its natural length and get a feel of the temperature by your lower lip again. It gets quite cool (Figure 7.1). Really surprised?

The high degree of elasticity of rubber bands can be drastically reduced by two external means: temperature and sunlight. Chill the same rubber band in the freezer for an hour or so. Can you stretch it as easily as before (be careful)? Probably, you cannot. In fact, it might even break when you stretch it. To demonstrate the effect of the sun takes a bit more time. Get a fresh rubber band and expose it to the sun for a week or so. Depending on the solar intensity of the place where you do the exposure, you may find that the rubber band becomes so brittle that it falls apart.

Rubber bands or elastomers* belong to a class of materials known as polymers. Polymers are made of long-chain molecules, each of which consists of many repeating units (*poly* = many, *mer* = unit). Naturally occurring polymers have been in use for a long time (e.g., rubber, cotton, wool, and silk).** Synthetic polymers became widely available shortly after World War II. Some of the properties observed in the preceding experiments are characteristic of polymers. In this chapter, we discuss how properties of polymers are related to their molecular structures.

7.2 POLYETHYLENE AS A TYPICAL POLYMER

Polyethylene is used in household containers, electrical insulation, chemical containers, tubings, housewares, packaging, prosthetic implants, etc. The chemical formula is $-CH_2-CH_2-CH_2-CH_2$ $-\ldots$, or $(-C_2H_4-)_n$, where n, the number of repeating C_2H_4 units, is known as the degree of polymerization. It can be synthesized via the reaction between hydrogen peroxide (H_2O_2) and ethylene (C_2H_4), in the presence of a catalyst such as palladium, as follows:

$$H_2O_2 \rightarrow 2OH$$

Initiation: $\quad OH + C_2H_4 \rightarrow HO - C_2H_4 -$

Propagation: $\quad HO - C_2H_4 - + C_2H_4 \rightarrow HO - C_2H_4 - C_2H_4 -$
$$\downarrow + C_2H_4$$
$$HO - C_2H_4 - C_2H_4 - C_2H_4 -$$
$$\downarrow + C_2H_4$$
$$HO - C_2H_4 - C_2H_4 - C_2H_4 - C_2H_4 -$$

---

* An elastomer is a polymer that can be elastically deformed to large strains.
** Considering that cell membranes, proteins, DNA, and RNA are made of long-chain molecules derived from simple units, one may argue that life is made of polymers.

FIGURE 7.1 Stretching and relaxing a rubber band.

TABLE 7.1
Boiling Points of Paraffins

Paraffin	Boiling Point (°C)
CH_4	−164
C_2H_6	−89
C_3H_8	−42
C_4H_{10}	−1
C_5H_{12}	36
C_6H_{14}	69

The propagation step occurs by first breaking one carbon–carbon double bond in ethylene (bond energy = 680 kJ/mol), followed by making two carbon–carbon single bonds (bond energy = 370 kJ/mol). This process results in a net energy release of 60 kJ/mol and is therefore energetically favorable. The chain propagation terminates due to the reaction between two similar chain fragments or between one chain fragment and the OH radical, as illustrated:

Termination 1: $HO - C_2H_4 - C_2H_4 - C_2H_4 - + - C_2H_4 - C_2H_4 - C_2H_4 - OH$
$\rightarrow HO - C_2H_4 - C_2H_4 - C_2H_4 - C_2H_4 - C_2H_4 - C_2H_4 - OH$
Termination 2: $HO - C_2H_4 - C_2H_4 - C_2H_4 - + - OH \rightarrow HO - C_2H_4 - C_2H_4 - C_2H_4 - OH$

Properties of polymers depend on the chain length or degree of polymerization n, so the trick in polymer synthesis is to learn how to control n. As an example, Table 7.1 shows the dependence of the boiling point of paraffins with general formula C_nH_{2n+2} on chain length. In the synthesis of polymers, we do not normally obtain polymers with a monodispersed distribution — that is, having just one value of n. Rather, there is a statistical distribution of n and hence of molecular weights. The average molecular weight is defined in two ways:

$$\overline{M_n} = \sum_i n_i M_i \quad \text{(number-averaged molecular weight)}, \qquad (7.1)$$

where n_i is the number fraction of polymers having chain length i and M_i is the molecular weight for polymers of chain length i. Another definition is:

$$\overline{M_w} = \sum_i w_i M_i \quad \text{(weight-averaged molecular weight)}, \qquad (7.2)$$

where w_i is the weight fraction of polymers having molecular weight M_i.

EXAMPLE

Consider that we have 10 polyethylene molecules with molecular weight distributions as follows:

Molecular Weight (MW)	No. of Molecules
2800	1
3360	1
3920	3
4480	3
5040	1
5600	1

Calculate $\overline{M_n}$ and $\overline{M_w}$.

Solution

Construct the following table to calculate the number fraction, total molecular weight, and weight fraction:

MW	No. of Molecules	No. Fraction	Total MW	Weight Fraction
2800	1	0.1	2800	0.067
3360	1	0.1	3360	0.080
3920	3	0.3	11760	0.281
4480	3	0.3	13440	0.321
5040	1	0.1	5040	0.120
5600	1	0.1	5600	0.134

- To calculate the number fraction (third column), divide the number of molecules with a given molecular weight (second column) by the total number of molecules (10).
- To calculate the total MW (fourth column), multiply the molecular weight (first column) by the number of molecules (second column). The molecular weight due to all 10 molecules is obtained by summing all entries in the total MW column and is equal to 42,000.
- To calculate the weight fraction (fifth column), divide the total MW (fourth column) by 42,000.

From Equation (7.1), the number-average molecular weight $\overline{M_n}$ is given by:

$$\overline{M_n} = 0.1 \times 2800 +$$

$$0.1 \times 3360 +$$

$$0.3 \times 3920 +$$

$$0.3 \times 4480 +$$

$$0.1 \times 5040 +$$

$$0.1 \times 5600 = 4200$$

$$\overline{M_w} = 0.067 \times 2800 +$$

$$0.08 \times 3360 +$$

$$0.281 \times 3920 +$$

$$0.321 \times 4480 +$$

$$0.12 \times 5040 +$$

$$0.134 \times 5600 = 4351$$

Why bother to define two different averages? In fundamental studies, it is easier to think in terms of how different *numbers* of polymer molecules affect the overall properties — for example, to obtain a polymer component with a certain elastic modulus. In order to synthesize such a polymer component, it is more convenient to think in terms of *weights* of individual polymer molecules, rather than *numbers* of molecules.

Question for discussion: In the preceding example, $\overline{M_w} > \overline{M_n}$. Is this always true?

Similarly, it is customary to have two definitions of the degree of polymerization: n_n, number-averaged degree of polymerization, and n_w, weight-averaged degree of polymerization, defined by:

$$n_n = \frac{\overline{M_n}}{m};\qquad\qquad\qquad (7.3a)$$

$$n_w = \frac{\overline{M_w}}{m},\qquad\qquad\qquad (7.3b)$$

where m is the molecular weight of the repeat unit (mer).

EXAMPLE

Using numbers from the preceding example, determine the number- and weight-averaged degrees of polymerization.

Solution

From the preceding example, the number-averaged molecular weight $\overline{M_n}$ is 4200, and the weight-averaged molecular weight $\overline{M_w}$ is 4351. For polyethylene, the molecular weight of the repeat unit $m = 28$. From Equation (7.3a), the number-averaged degree of polymerization n_n is given by:

$$n_n = \frac{\overline{M_n}}{m} = \frac{4200}{28} = 150.$$

From Equation (7.3b), the weight-averaged degree of polymerization n_w is given by:

$$n_w = \frac{\overline{M_w}}{m} = \frac{4351}{28} = 155.$$

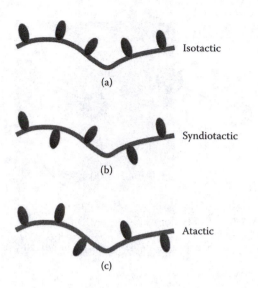

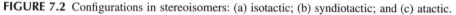

FIGURE 7.2 Configurations in stereoisomers: (a) isotactic; (b) syndiotactic; and (c) atactic.

7.3 BEYOND POLYETHYLENE: POLYMER STRUCTURES

7.3.1 STEREOISOMERS

The field of polymers offers numerous opportunities for materials synthesis. Starting with the basic structure of polyethylene $(-C_2H_4-)_n$, any of the hydrogen atoms can be substituted with other elements or groups of atoms. For example, replacing one of the hydrogen atoms with chlorine makes polyvinyl chloride, with a benzene ring it makes polystyrene, with $-CN$ it makes polyacrylonitrile, etc. All these polymers clearly have different properties. In addition, for a given polymer, the replacement atoms or groups of atoms may be on the same side of the chain (isotactic configuration), they may alternate (syndiotactic), or they may position randomly (atactic) (Figure 7.2). In spite of having identical compositions, these different forms of the polymer may have different properties and are known as *stereoisomers*.

7.3.2 LINEAR POLYMERS

Examples given in the preceding paragraph are known as *linear polymers*. The main backbone of *polyethylene* consists of a linear and flexible chain of single-bonded carbon atoms.* We can understand this flexibility by looking at a section of the polyethylene chain (Figure 7.3). As indicated, the lightly shaded carbon atom can be at any position along the dotted circle, as long as the C–C–C bond angle is maintained at 109.5°. As a result, a linear polymer with a single-bonded carbon backbone with small side groups such as polyethylene tends to twist and coil, much like spaghetti (Figure 7.4).** The average length L of polyethylene $(-C_2H_4-)_n$ is given by:

$$L = d\sqrt{2n} \, , \tag{7.4}$$

where d is the carbon–carbon bond length in polyethylene (0.154 nm). The factor 2 comes from the fact that each ethylene unit has two carbon–carbon bonds.

---

* When stretched, the maximum length of polyethylene $(-C_2H_4-)_n$ is equal to $(2n) \, d \, \sin(109.5°/2)$, where d is 0.154 nm, the carbon–carbon single bond length.

** In contrast, a carbon–carbon double or triple bond is rotationally rigid because of the charge distribution of π bonds. Also, large side groups (e.g., benzene) tend to inhibit rotation.

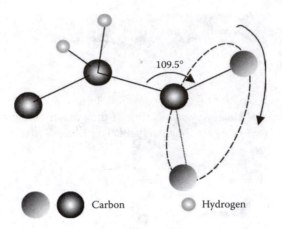

FIGURE 7.3 Rotation about the carbon–carbon single-bond axis in polyethylene.

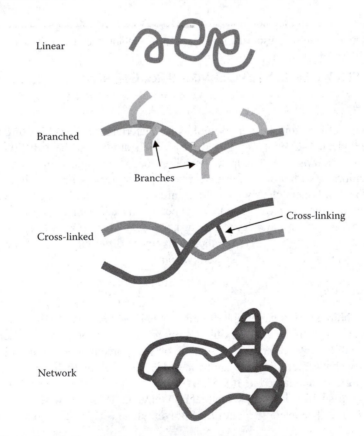

FIGURE 7.4 Linear, branched, and network polymers.

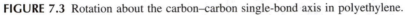

Derivation of Equation (7.4)

The length of the polymer chain can be represented by a vector L, composed of individual C–H bond segments, s_i, where $i = 1$ to $2n$:

$$\vec{L} = \vec{s} + \vec{s}_2 + \ldots + \overline{s_{2n}},$$

so that,

$$L^2 = \vec{L}.\vec{L} = \sum_{i=1}^{2n} \vec{s_i}^2 + \sum_{i>j} \vec{s_i}.\vec{s_j}.$$

The first term is equal to $2nd^2$, where d is the absolute value for the vector s_i. The average value of the second term is equal to zero because of the random orientation of one C–C bond relative to another — that is, $s_i . s_j$ is equally likely to be positive as it is to be negative. Therefore, we have:

$$|L| = \sqrt{average(L^2)}$$

$$= \sqrt{2nd^2}$$

$$= d\sqrt{2n}.$$

7.3.3 BRANCHED POLYMERS

There is no restriction in terms of what we can use to replace the hydrogen atoms. We can replace them with a long chain of atoms. In this case, we have a *branched polymer*. The existence of side branches makes it difficult to pack the polymer molecules tightly, thus resulting in a lower mass density for the polymer material. These side branches tend to inhibit sliding of the main chains past each other, thus increasing the mechanical strength. This is one of several strategies to improve mechanical properties of polymers.

7.3.4 CROSS-LINKED POLYMERS

Through a chemical reagent or ultraviolet light, two or more polymer chains can be covalently linked together. The resulting *cross-linked polymers* are stronger and more rigid. One example of cross-linking is the making of rubber tires, in which polymer chains are connected via sulfur linkages. The process is known as vulcanization. The amount of sulfur used in vulcanization is typically on the order of a few weight percent. Although the strength can be increased with increasing sulfur concentration, the resulting increase in cross-linking also makes the rubber brittle.* In our rubber band experiments, ultraviolet light from the sunlight produces cross-linking — extended sunlight exposure makes rubber bands brittle.

Since oxygen is just above sulfur in the periodic table, can oxygen be used to link molecules? The answer is yes. Common examples of oxygen linkage do not appear in structural polymers, but rather in lubricants and biochemicals. For example, a polyphenyl fluid as shown in Figure 7.5 is used as a lubricant for the SR-71, a supersonic spy plane developed during the Cold War era. It consists of several benzene rings (between 2 and 10) linked by oxygen atoms. The large molecular weight ensures low vapor pressure at elevated temperatures. Because of the resonance energy associated with the benzene ring, such a structure is stable against high temperature and oxidation. Finally, the oxygen linkage allows molecular rotation about the bond axis, resulting in a flexible molecule. This means that a polyphenyl compound can remain as a liquid at low temperatures, providing the optimum combination of viscosity and thermal and oxidative stability needed for high-temperature engine operation of the SR-71.

---

* That is why burning rubber tires produces such a foul smell and are an environmental hazard. The sulfur in the rubber tires turns mostly into sulfur dioxide, which combines with water vapor in the lung to form sulfurous acid (H_2SO_3).

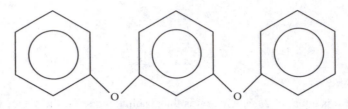

FIGURE 7.5 Chemical structure of a polyphenyl lubricant used in the SR-71.

Another example where oxygen linkage occurs is table sugar (sucrose). Sucrose is formed by the condensation of two simple sugars (monosaccharides), glucose and fructose, followed by the elimination of water, as shown in Figure 7.6(a). The resulting oxygen linkage between these monosaccharides is known as a glucosidic bond. The term polysaccharide refers to a polymer chain of monosaccharide units linked together via these glucosidic bonds (e.g., complex sugars and cellulose). It is interesting to note that one type of artificial sweetener, sucralose, is related to sucrose by substituting three hydroxyl groups with chlorine, as shown in Figure 7.6(b). In spite of the similarity between sucrose and sucralose, the presence of carbon–chlorine bonds makes the body unable to metabolize sucralose, hence its application as a zero-calorie sugar substitute.

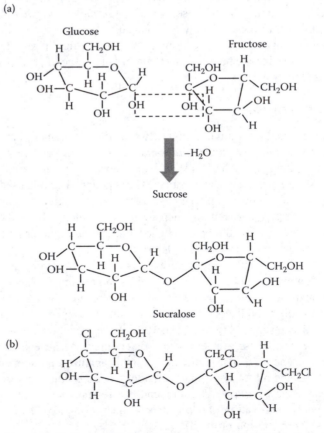

FIGURE 7.6 (a) Condensation of glucose and fructose to form sucrose; (b) chemical structure of sucralose, a zero-calorie sugar substitute.

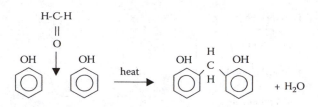

FIGURE 7.7 Condensation reaction between phenol and formaldehyde.

7.3.5 Network Polymers

In the synthesis of polyethylene, the carbon–carbon double bond of ethylene, the basic unit of the polymer, is opened up so that it has two active sites for the polymerization reaction. This results in the formation of a linear polymer. On the other hand, if the mer unit has three or more active sites for the polymerization reaction, a three-dimensional network can be formed. The result is a *network polymer*. These polymers are strong and rigid.

Bakelite (technical name: poly-phenol-formaldehyde) is one such polymer. It is synthesized via a condensation reaction between phenol and formaldehyde (Figure 7.7). The reaction results in the bridging between two phenol molecules and can occur at any of the five positions in the phenol molecule. As the condensation reaction progresses, the polymer builds in three dimensions.

Since network polymers are rigid after they are formed, it is necessary to synthesize objects made of these polymers in a special way. The required reactants are first mixed together at room temperature and packed inside a mold of the desired shape. The temperature is then raised to turn on the polymerization reaction (a catalyst is normally required to speed up the reaction). The polymer is formed and set into a fixed shape defined by the mold. These polymers are therefore known as *thermosets* or *thermoset plastics*. Once formed, they cannot be melted and remolded into another shape. That is why thermoset polymers are significantly more difficult to recycle than polymers such as polyethylene, which can be remelted by heating and shaped repeatedly using injection molding or extrusion. The latter polymers are known as *thermoplastics*.

7.4 COMMON POLYMERS AND TYPICAL APPLICATIONS

Figure 7.8 shows the chemical structures of common polymers. Typical applications for selected polymers are:

Polyvinyl chloride (PVC): electrical insulation pipes, siding, upholstery, floor covering
Polyacrylonitrile: fibers for clothing and blankets
Teflon (polytetrafluoroethylene, PTFE): nonstick coatings, seals, chemical containers and pipes, cable insulation
Polymethylmethacrylate (PMMA): plexiglass windshields for aircraft, outside signs, safety shields and goggles, dental materials
Polycarbonate: safety shields, helmets, prescription lens, aircraft components, boat propellers
Bakelite (phenol-formaldehyde): electrical connectors and switches, motor and telephone housing, billiard balls

7.5 SOLID SOLUTIONS (COPOLYMERS)

As noted earlier, the field of polymers offers numerous opportunities for materials synthesis. The possibility of achieving new properties by mixing polymers is almost unlimited. Recall our discussion on the Fe–C phase diagram. By introducing the right amount of carbon into iron and with the

Polyethylene $-C_2H_4-$

Teflon(PTFE) $-C_2F_4-$

Polyvinyl chloride $-CH_2-CHCl-$

Polyacrylonitrile $-CH_2-CHN-$

Polystyrene $-CH_2-CH-$

Polypropylene $-CH_2-CHCH_3-$

Polymethylmethacrylate (PMMA)

$$-C-CH_2-\underset{\underset{COO-CH_3}{|}}{\overset{\overset{CH_3}{|}}{C}}-$$

Polycarbonate

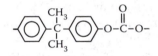

Polyester (polyethylene terphthalate PET)

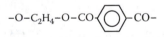

Bakelite

Nylon $-NH-(CH_2)_6-NH-CO-(CH_2)_4-CO-$

Aramid (Kevlar 49)

FIGURE 7.8 Chemical structures of several common polymers.

proper heat treatment, a whole series of steels with different combinations of strength and toughness can be obtained. There is every reason to believe that the same can be done with polymers.

A copolymer can be thought of as a solid solution of mer units, with properties that are not given by the rule of mixtures. Consider a copolymer consisting of two different mer units, A and B. These mer units can be arranged in four different ways (Figure 7.9):

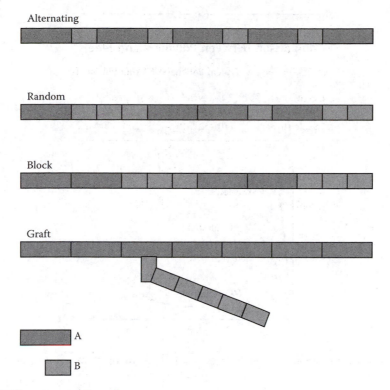

FIGURE 7.9 Different types of copolymers.

- Alternating: The A and B units repeat in an alternating (ABAB...) fashion.
- Random: The A and B units are distributed randomly along the polymer chain.
- Block: The A and B units appear as blocks along the chain (AAAABBBB...).
- Graft: The A units make up the main chain of the polymer, while blocks of B units are grafted to or form side branches along the main chain.

For example, modern automobile tires are made of random copolymers of butadiene $-CH_2-CH=CH-CH_2-$ and styrene $-CH_2-C(C_6H_5)H-$ or butadiene and acrylonitrile $-CH_2-CHN-$. As will be discussed in the section on mechanical properties, the addition of styrene or acrylonitrile inhibits the sliding of chains relative to one another, thus improving rigidity of the rubber.

7.6 CRYSTALLINITY

Most metals are crystalline, while most polymers are semicrystalline or amorphous. The ability of polymers to form crystalline structures depends on two factors: cooling rate and molecular structure. Because polymer chains are usually long and bulky, it takes time for polymer molecules to diffuse and pack in an ordered structure. If the cooling is too fast, there may not be sufficient time for diffusion to occur before the sample temperature becomes too low for further molecular motion. Generally, crystalline growth is not favorable for polymers with complex mer structures or with bulky side groups, such as branched polymers, network polymers, atactic polymers, or random copolymers. This is due to the difficulty of packing complex and randomly shaped molecules in perfect crystalline order. Crystalline polymers, as a general rule, have higher density, are stronger, and are more resistant to heat or chemical attack.

TABLE 7.2
Comparison between Polymers and Steels

	Typical Polymers	Common Steels
Density	1–2 g/cm^3	8 g/cm^3
Modulus	0.5–5 GPa	200 GPa
Tensile strength	10–50 MPa	200–800 MPa

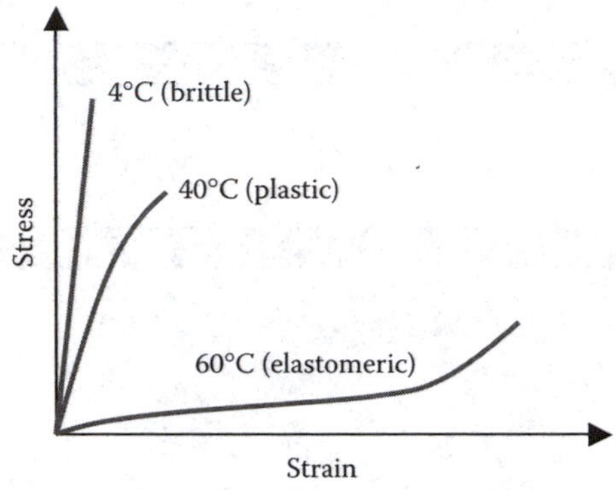

FIGURE 7.10 Schematic diagram showing the stress–strain response of PMMA at different temperatures.

7.7 MECHANICAL PROPERTIES

Table 7.2 shows the comparison between typical polymers and common steels. By common experience, most polymers are weaker than metallic alloys. However, some polymers have mechanical properties comparable to those of steels and aluminum alloys. For example, aramid (Kevlar 49) has an elastic modulus of 130 GPa and density of only 1.3 g/cm^3 (about one sixth of steels), compared with the elastic modulus of about 200 GPa in steels and about 70 GPa in typical aluminum alloys.

Polymers can exhibit elastomeric (large elastic strain), plastic, and brittle mechanical behavior, depending on their structure and testing temperature. Figure 7.10 shows schematically the stress–strain response of PMMA at different temperatures. At 60°C, PMMA behaves like rubber bands, capable of sustaining large elastic strains. (Note that elastic deformation means reversible deformation and is not necessarily equivalent to having a linear stress–strain curve.) At lower temperatures (40°C), plastic deformation occurs. At even lower temperatures (4°C), PMMA is brittle like glass. The influence of temperature on the mechanical properties is due to the thermally activated motion of polymer chains. As discussed later, below a certain temperature (glass transition temperature), polymers behave like glass. Note that the definition of 0.2% offset yield strength used for metals does not apply to polymers.

7.7.1 DEFORMATION MECHANISMS OF SEMICRYSTALLINE POLYMERS

Many semicrystalline polymers have the lamellar structure shown in Figure 7.11 — that is, lamellar chain-folded crystallites linked to each other by tie chains in amorphous layers. Mechanical properties of semicrystalline polymers depend on the properties of crystalline and amorphous phases.

7.7.1.1 Elastic Deformation

Elastic deformation is due mainly to the extension of the backbone chain by bending and stretching of covalent bonds, relative displacement of chains, and uncoiling of the polymer chain. In

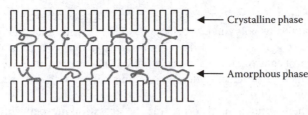

FIGURE 7.11 Schematic diagram showing the lamellar structure in a semicrystalline polymer.

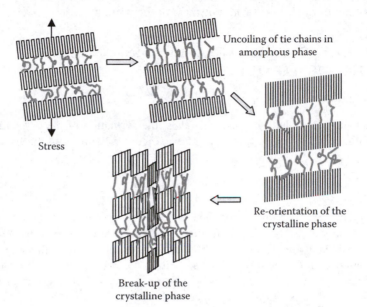

FIGURE 7.12 Elastic and plastic deformation of a semicrystalline polymer.

semicrystalline polymers, the elastic strain comes from the deformation and uncoiling of polymer chains in the amorphous phase.

7.7.1.2 Plastic Deformation

As shown in Figure 7.12, the plastic deformation of a semicrystalline polymer goes through several stages, depending on the strain involved. The initial plastic deformation is due to the uncoiling and sliding of tie chains in the amorphous phase. With increasing stress, tilting of the lamellar crystalline layers to align with the stress axis begins to occur. This is then followed by sliding and separation of lamellar crystalline segments.

7.7.2 STRENGTHENING STRATEGIES

There are several strategies to increase the modulus and strength of polymers:

- Increase crystallinity to give tighter packing and hence stronger intermolecular bonding between chains.
- Attach side groups to interfere with chain sliding. For example, the tensile modulus of polyethylene is ~0.4–1.1 GPa, increasing to 1.1–1.5 GPa for polypropylene (with methyl side groups) and 2.8–3.5 GPa for polystyrene (with benzene ring side groups).

- Attach polar atoms to increase attractive interactions between polymer chains. For example, the tensile modulus of polyvinyl chloride is 1.4 GPa (greater than that of polyethylene), and its tensile strength is 40–80 MPa, compared to 15–35 MPa for polyethylene.
- Introduce electronegative atoms such as oxygen and nitrogen in the main backbone of the polymer chain to promote hydrogen bonding between chains (e.g., aramid, nylon, and polycarboxylamide).
- Cross-link. When sulfur is used to provide the cross-linking between polymer chains, the process is known as vulcanization. Moderate cross-linking is used to impart elastomeric properties to polymers (i.e., ability to sustain large elastic strains). High degrees of cross-linking result in network polymers or thermosets.

7.8 CRYSTALLIZATION, MELTING, AND GLASS TRANSITION TEMPERATURES

7.8.1 CRYSTALLIZATION

When a polymer melt is cooled to produce crystals, the formation of the crystalline phase takes time. The crystallized fraction as a function of time (Figure 7.13) follows the Avrami Equation:

$$y = 1 - \exp(-kt^n),\tag{7.5}$$

where y is the crystallized fraction, t is the time, and k and n are constants for a given temperature. Note that for most polymers, 100% crystallization is not possible because of interference from side groups. In this case, y only applies to the portion of the polymer sample that can be crystallized. Since crystallization of polymer molecules takes time, the crystallized fraction depends on the cooling rate. If the cooling is done too rapidly, polymer molecules may not have enough time to move into ideal crystalline positions, and the crystallized fraction will be small.

7.8.2 MELTING

Unlike metals and ceramics, polymers melt over a range of temperatures for two reasons. First, as discussed in the preceding paragraph, depending on how fast the polymer sample is cooled from the melt, the crystallized fraction may vary. Crystalline and amorphous polymers do not have the same melting temperature. Second, a typical polymer sample consists of a certain molecular weight distribution, which gives rise to a range of melting temperatures.

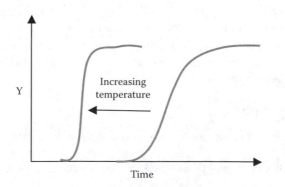

FIGURE 7.13 Crystallized fraction versus time at different temperatures. As expected, it takes a shorter time for crystallization to occur at higher temperatures.

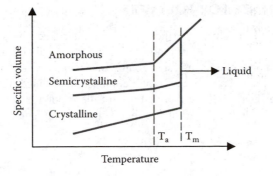

FIGURE 7.14 Specific volume versus temperature for different types of polymers.

7.8.3 GLASS TRANSITION TEMPERATURE

Below a certain temperature T_g (glass transition temperature), thermal energy is insufficient to activate the motion of large chain segments. As a result, polymers below the glass transition temperature T_g are rigid and brittle. Above T_g, they become rubbery, are easily deformable, and have improved fracture toughness. The variation of specific volume (volume per mole or per unit mass) versus temperature for different types of polymers (amorphous or glassy, semicrystalline, and crystalline) is shown in Figure 7.14. Typically, T_g is about 0.5–0.8 times the melting point T_m. Generally, any polymer modification or processing that strengthens inter- or intramolecular interaction raises T_g and T_m.

An amorphous polymer behaves like glass (supercooled liquid) below T_g, a rubbery solid above T_g, and a viscous liquid as T_m is approached. Below the glass transition temperature, motion of polymer chains is restricted, and creep is negligible. At temperatures between T_m and T_g, a combination of elastic deformation and viscous flow of one chain past another (viscoelastic behavior) can be observed.*

7.9 RUBBER BAND MYSTERY UNVEILED

In the unstressed state, an elastomer consists of coiled chains and is disordered. When it is stretched, the chains uncoil, and the system becomes more ordered (low entropy). When the rubber band is relaxed to the natural state, the polymer chains coil and twist, returning to a disordered (high-entropy) state. Given an entropy increase of ΔS** when we relax the rubber band, the amount of heat absorbed by the rubber band is equal to $T\Delta S$.*** It is this heat absorption that gives us the cool sensation when we relax the rubber band.

Question for discussion: Without applying any external force, what will happen to the length of the rubber band when we warm it to slightly above room temperature? (Hint: When given thermal activation, will the polymer chains become straighter or more "wiggled"?)

---

* The consequence of this viscoelastic behavior is that when a load is applied, the resulting strain has two components. The first, normal elastic strain, occurs instantaneously. The second component, due to viscous flow, increases with time.

** Statistical thermodynamics analysis shows that the entropy of a polymer chain with N links, each of length a, is equal to $k_B (\Omega_o - L^2/2Na^2)$, where k_B is the Boltzmann constant, L the head-to-tail length of the polymer, and Ω_o some constant. Therefore, stretching the polymer (increasing L) decreases the entropy and vice versa.

*** The same heat absorption phenomenon due to entropy increase occurs around us: melting of ice to liquid water at 0°C (which absorbs about 80 cal/g), vaporization of water to steam at 100°C (which absorbs about 540 cal/g), or transition from a superconductor to a normal conductor.

7.10 FIRE RETARDANTS FOR POLYMERS

Polymers are widely used in fabrics, textiles (e.g., polyesters, acrylics, and nylons), and foams (e.g., polyurethane). Once ignited, a typical polymer can burn rapidly because carbon and hydrogen atoms present in most polymers act as fuel. Foams made of polyurethane (e.g., airline seats or mattresses) are especially dangerous because their combustion produces toxic cyanide fumes. Many tragedies have occurred because of the flammability of these polymer products. As a result, much research has gone into the development of fire retardants for polymers.

EXAMPLE

Calculate the heat released (enthalpy) when 1 mole of octadecane ($C_{18}H_{38}$) is combusted in oxygen to produce water vapor and carbon dioxide under standard temperature and pressure conditions. It is given that the standard enthalpy is –567 kJ/mol for octadecane, –394 kJ/mol for carbon dioxide, and –242 kJ/mol for water vapor.

Solution

The chemical reaction is given by:

$$C_{18}H_{38} + 27.5O_2 \rightarrow 18CO_2 + 19H_2O.$$

Therefore, the enthalpy change is equal to:

$$18(-394) + 19(-242) - (-567) = -11,123 \text{ kJ/mol of octadecane.}$$

One mole of octadecane weighs 254 g. This means that combustion of 1 g of octadecane produces about 44 kJ, typical of hydrocarbons.

Fire can only be sustained when all three components (heat, fuel, and oxygen) are present. The primary function of a fire retardant is to interfere with the presence of one or more of these components. For example, tribromoneopentyl alcohol (TBNPA) is an effective fire retardant (Figure 7.15) for polyurethanes. It melts at a relatively low temperature (65°C), absorbing significant thermal energy during the process. At higher temperatures, TBNPA decomposes, absorbing even more thermal energy. More important, the decomposition pro-

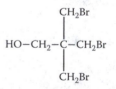

FIGURE 7.15 Tribromoneopentyl alcohol, containing 73 w/o bromine.

duces a blanket of bromine (which is heavier than air) that cuts off oxygen access to the burning site.

A nonhalogen alternative is melamine or melamine derivatives (i.e., salts formed with acids such as boric, phosphoric, and cyanuric acids).* Melamine melts at 354°C, but it begins to sublime just below 200°C, absorbing 120 kJ/mol in the process. The sublimation not only absorbs heat, but also produces a vapor that dilutes the fuel–air mixture. At higher temperatures, the vapor decomposes, absorbing about 1950 kJ/mol. One decomposition product is nitrogen (Figure 7.16), which further helps to smother the fire. With their low solubility in water, melamine and its derivatives impregnated into fabric and textile products can withstand repeated washing.

---

* The reason for avoiding bromine is obvious. Apart from having a pungent smell, bromine dissolves in water (present in mucous membranes in nasal passages and the lungs) to form HBr, which can be highly irritating.

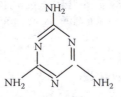

FIGURE 7.16 Melamine (2,4,6 triamino-1,3,5 triazine).

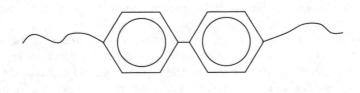

Example: $CH_3-CH_2-CH_2-CH_2-CH_2$—⬡—⬡—CN

FIGURE 7.17 General structure of a liquid crystal molecule.

7.11 SELECTED ELECTRO-OPTICAL APPLICATIONS

Because of their transparency, polycarbonates are often used in optical applications. The index of refraction at visible wavelengths is close to that of glass (~1.5) so that polycarbonates can directly replace glass, as in prescription lenses, aircraft windows, laminates for car windshields, etc. Major advantages of polycarbonates over glass include better toughness and lighter weight.

Normally, when a solid turns into a liquid, any long-range order disappears. Some polymers, however, retain some degree of long-range order or crystallinity above the melting point. These are known as liquid crystals or, more specifically, thermotropic liquid crystals.* The general structure of a liquid crystal is a central nucleus made of two linked benzene rings and two chains along the axis of the molecule, as shown in Figure 7.17. Consider one such example: pentyl cyanobiphenyl C_5H_{11}–Ph–Ph–CN, where Ph represents the phenyl group. Since N is more electronegative than C, N acquires a small negative charge. As a result, application of an electric field tends to align this molecule along the field direction. Such alignment affects the scattering of light. This phenomenon is the basis of the liquid crystal display.

Also, aligning these molecules with an electric field creates a material with a noncentrosymmetric structure (i.e., one without mirror symmetry). Such materials exhibit interesting nonlinear optical effects such as sum frequency generation (i.e., generation of light with frequency $v_1 + v_2$ when illuminated by light with frequency v_1 and v_2). This property allows design of optoelectronic circuits for high-speed data transmission and optical communications.

An emerging application of polymers is in lithium ion batteries and fuel cells. In both cases, a lightweight and robust material is needed as an electrolyte to transport ions between electrodes. In lithium ion batteries, polyethylene oxide (PEO) and its variants appear to be the material of choice to date. The structure of PEO is $(-O-CH_2-CH_2-)_n$. There are at least two factors that make PEO an appropriate electrolyte for positive ions. First, oxygen is more electronegative than carbon,

---

* Lyotropic liquid crystals acquire long-range order due to their interaction with solvent molecules. For example, stearic acid molecules align themselves on the water surface because of the attractive interaction between the carboxyl groups and water molecules.

so it has a slightly negative charge. As a result, oxygen sites become anchor points for positive ions. Second, the distance between oxygen sites is not too great, which makes it easier for the positive ions to hop from one site to another, thus improving the ionic conductivity of the electrolyte.

7.12 POLYMER AND LIFE SCIENCES

The rich variety of polymers can be awe inspiring. As we discuss throughout this chapter, polymers can occur in many structural forms, which directly affect their chemical, mechanical, and physical properties. It is no wonder that some polymers are stronger than steel, some more electrically conducting than metals (some polymers are superconductors), and some with dielectric strength comparable to the best insulating ceramics.

In our discussion of copolymers, we talk about different arrangements of mer units. Here, we can see the parallel between polymer science and life science. A protein molecule consists of a string of amino acids (there are 20 such amino acids). Therefore, we can consider proteins to be polymers. Properties and functions of a protein molecule depend not only on its composition, but also on its shape or conformation, which in turn depends on the amino acid sequence, analogous to polymer properties depending on the arrangement of individual mer units.

If we choose the individual mer units to be one of those well-known nucleic acids in biochemistry — adenosine, cytosine, guanine, thymine, and uracil, then we can create copolymers that mimic segments of DNA or RNA. Such segments may be useful for biomolecular synthesis, detection of biomolecules and pathogens, and medical diagnostics.

Question for discussion: The main linkage between mer units in nylon and Kevlar chains has the following structure:

This is the same linkage between amino acids in a protein molecule and is called a peptide bond in biochemistry. A protein molecule folds into a specific shape or conformation due to hydrogen bonding between CO and NH segments. On the other hand, nylon and Kevlar chains do not fold into any specific shapes. Discuss why there is such a difference.

PROBLEMS

1. Consider a polyethylene molecule $(-C_2H_4-)_n$ with $n = 50$.
   a. What is the root-mean-square length (coiled length) of this molecule?
   b. What is the length of this molecule when stretched? Note that the C–C–C angle is 109.5°.
   c. When a polymer is suddenly released from the stretched configuration, what will happen to its temperature? Why?
2. Study the figure in the text showing the structure of an engineering plastic known as aramid (Kevlar 49). The Young's modulus of aramid is among the highest of all polymers (131 GPa, cf. ~200 GPa for steels). Since the density of aramid is only one sixth of steel, aramid is pound for pound almost four times stronger than steel. Explain why this material has such a high Young's modulus.
3. Polyethylene (PE) with 100% crystallinity has a density of 1.01 g/cm^3, while amorphous PE has a density of 0.90 g/cm^3. The density of high-density polyethylene (HDPE) is 0.96 g/cm^3. Show that the volume fraction of crystalline PE in HDPE is about 55%.

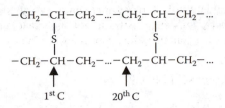

FIGURE 7.18 Sulfur cross-linking of two polyethylene chains.

4. Vulcanization
   a. Consider two polyethylene chains cross-linked by sulfur atoms every 20 carbon atoms, as shown in Figure 7.18. Show that the weight fraction of sulfur in this cross-linked polymer is about 5%.
   b. Rarely is more than a few weight percent of sulfur added in vulcanization to make polymers stronger. Why?
5. A 100-g sample of synthetic rubber, polybutadiene $(-CH_2-CH=CH-CH_2-)_n$ cross-linked with sulfur, is found to contain 3 g of sulfur. Assuming that there is one sulfur atom per connection, calculate the average number of connections per butadiene monomer pair.
6. The elastic moduli for three polymers as shown in Figure 7.19 are ranked in an ascending order:

$$E(a) < E(b) < E(c)$$

   a. Explain why polymer (b) is stiffer than (a) (i.e., has a larger elastic modulus).
   b. Explain why polymer (c) is stiffer than (b).
   c. Polymer (c) has the highest stiffness of the three polymers, but it has a major drawback. What is it? How can this drawback be minimized or eliminated? (Hint: the sulfur concentration is about 37 w/o.)
7. When a polymer is mechanically deformed above the glass transition temperature, elastic and plastic deformation can occur. Using sketches where appropriate, describe briefly any two major deformation mechanisms in polymers.
8. Let us estimate the temperature decrease when a polymer chain is relaxed, as demonstrated in the rubber band experiment. The entropy of a polymer chain with N links, each of length a, is equal to k_B $(\Omega_o - L^2/2Na^2)$, where k_B is the Boltzmann constant, L the head-to-tail length of the polymer, and Ω_o some constant.

$(-CH_2-CH_2-)_n$

(a)

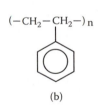

(b)

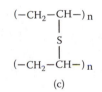

(c)

FIGURE 7.19 Ranking the elastic modulus of three polymers.

   a. Show that the entropy change dS is equal to $-(k_B L/Na^2)\, dL$.
   b. Heat energy is absorbed from the surrounding (your lip in the rubber band experiment) to obtain this entropy change such that $-T\, dS = C\, dT$, where T is the absolute temperature and C the heat capacity of the polymer chain. Let us assume that the heat capacity of the polymer chain is equal to $3Nk_B$. Show that:

$$3Nk_B.dT = \frac{k_B L}{Na^2} TdL.$$

c. The average natural length L_o of a polymer chain is $a\sqrt{N}$. Show that when the polymer chain at temperature T_i is relaxed from a stretched length of $3L_o$ to L_o, the final temperature T_f is given by:

$$\ln\frac{T_f}{T_i} = -\frac{4}{3N}.$$

d. For $N = 40$ at an initial temperature of 300 K, show that the temperature drop is about 10 K.

9. Certain plastic containers and items are labeled as "dishwasher safe." Discuss what this means along two lines:
   a. What will happen to "unsafe" plastic containers after they are washed?
   b. What is the specific property that makes a polymer "dishwasher safe"?

10. Consider three compounds as shown in Figure 7.20: naphthalene, diphenyl ether, and diphenyl thioether, with the following melting points: 81, 28, and –40°C. Discuss why the melting point decreases in the order as shown, in terms of the flexibility of the molecule and attractive interactions between molecules. Note that oxygen is more electronegative than sulfur.

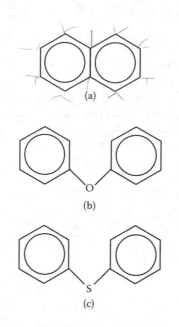

FIGURE 7.20 Chemical structure of (a) naphthalene; (b) diphenyl ether; and (c) diphenyl thioether.

8 Corrosion and Oxidation of Metals and Alloys

8.1 SILVERWARE CLEANING MAGIC

One unpleasant household chore is to clean off the tarnish on silverware. The usual method is to use a soft cloth to rub and buff the silverware. That takes muscle and patience. Fortunately, there is a better way (Figure 8.1).* Take a quart of warm water. Dissolve into it one tablespoon of water softener or washing soda and one tablespoon of salt. Lay a sheet of aluminum foil at the bottom of the water container. Now, here is the magic: immerse the silverware into the water *and* make sure that it is in contact with the aluminum foil. In seconds, the tarnish is gone. Remove the silverware. Rinse in tap water and dry and it is done!

Corrosion is defined as the deterioration of materials due to chemical attack by the environment. The process involves the flow of electrical currents between two surface locations (cathode and anode) at different electrochemical potentials. Corrosion can be undesirable: It results in the loss of materials and infrastructure (e.g., rusting of car bodies; corrosion of aircraft structures, roads, and bridges). It can also be desirable: It is exploited in the generation of electrical power in batteries and fuel cells, electroplating, chemical etching in the manufacturing of integrated circuits, and, of course, the preceding example of cleaning silverware.

8.2 CONVENTIONAL EXAMPLE OF CORROSION

We will return to the silverware cleaning magic a bit later. For the moment, let us consider a more conventional example of corrosion: the dissolution of zinc in hydrochloric acid, as represented by the standard equation: $Zn + 2HCl \rightarrow ZnCl_2 + H_2$. Alternatively, we can represent this reaction as two half-cell reactions. The first one is:

$$Zn \rightarrow Zn^{2+} + 2e^-. \qquad (8.1a)$$

This is an oxidation reaction. The site at which oxidation takes place is called the anode.** In this case, the Zn ion goes into solution (Figure 8.2).

In other situations, the Zn ion may form a deposit or insoluble precipitate by reaction with some anion in the solution. The second half-cell reaction is:

$$2H^+ + 2e^- \rightarrow H_2. \qquad (8.1b)$$

This is a reduction reaction. The site at which reduction takes place is called the cathode.*** In this case, H^+ ions in solution accept electrons from zinc to produce atomic hydrogen. Hydrogen atoms then combine to form hydrogen gas. The total reaction is represented schematically in Figure 8.2. The anode and cathode are at different locations on the same Zn surface.

---

* I learned this recipe from a book called *Cleaning Hints* by Graham and Rosemary Haley.
** Memory aid: Oxidation and anode both begin with a vowel.
*** Memory aid: Reduction and cathode both begin with a consonant. Another useful memory aid is OILRIG — Oxidation Is Loss (of electrons), Reduction Is Gain (of electrons).

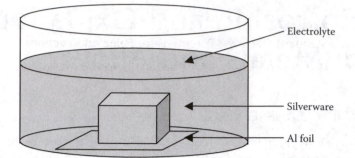

FIGURE 8.1 Cleaning silverware.

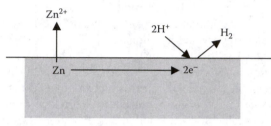

FIGURE 8.2 Dissolution of zinc in hydrochloric acid.

8.3 ELECTRODE POTENTIALS

Why is it not possible for the opposite — hydrogen giving up electrons, which are then taken up by Zn^{2+} resulting in the deposition of Zn — to occur? The answer lies in the energetics of an element giving up electrons. The tendency of a metal to give up electrons such as shown in Equation (8.1a) is generally referenced to that of hydrogen ($H_2 \rightarrow 2H^+ + 2e^-$) using an arrangement shown in Figure 8.3 for the case of zinc. The standard zinc electrode is zinc in contact with a solution containing $1\ M\ Zn^{2+}$ ions at 25°C ($1\ M\ ZnCl_2$ as illustrated in Figure 8.3). The standard hydrogen reference electrode is hydrogen gas at 1 atm bubbling through a solution containing $1\ M\ H^+$ ions at 25°C ($1\ M\ HCl$ as illustrated in Figure 8.3).

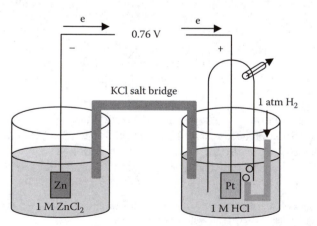

FIGURE 8.3 Measurement of the standard electrode potential of zinc relative to a standard hydrogen electrode, all at 25°C.

TABLE 8.1
Standard Electrode Potentials for Selected Reactions

Electrode Reaction	Standard Electrode Potential (volt)
$OH^- \rightarrow OH + e^-$	+2.02
$Au \rightarrow Au^{3+} + 3e^-$	+1.50
$2Cl^- \rightarrow Cl_2(g) + 2e^-$	+1.36
$2H_2O \rightarrow O_2 + 4H^+ + 4e^-$	+1.23
$Ag \rightarrow Ag^+ + e^-$	+0.80
$4OH^- \rightarrow O_2 + 2H_2O + 4e^-$	+0.40
$Cu \rightarrow Cu^{2+} + 2e^-$	+0.34
$H_2 \rightarrow 2H^+ + 2e^-$	+0.00
$Co \rightarrow Co^{2+} + 2e^-$	−0.28
$Fe \rightarrow Fe^{2+} + 2e^-$	−0.44
$Cr \rightarrow Cr^{3+} + 3e^-$	−0.74
$Zn \rightarrow Zn^{2+} + 2e^-$	−0.76
$Al \rightarrow Al^{3+} + 3e^-$	−1.66
$Na \rightarrow Na^+ + e^-$	−2.71
$Li \rightarrow Li^+ + e^-$	−2.96

More likely to occur

The platinum electrode in the hydrogen reference electrode serves two functions. First, it acts to maintain equilibrium between H_2 and H^+. Second, it provides an electrical contact to the hydrogen half-cell. The two half-cells are typically connected by a KCl gel bridge, which allows ionic currents to pass through, but not the intermixing of the two solutions. In this standard configuration, the potential difference is measured to be 0.76 V, negative on the Zn side. The standard electrode potential for Zn is then designated as −0.76 V.

Table 8.1 gives standard electrode potentials for selected reactions. These electrode potentials are measured with respect to the standard hydrogen electrode under open-circuit conditions: 1 atm hydrogen bubbling in 1 M HCl, with the metal in contact with solution containing 1 M metal ions at 25°C. Half-cell reactions with more negative electrode potentials are more likely to occur. In the silverware experiment where we have silver and aluminum side by side in an electrochemical environment, aluminum will be oxidized (with standard potential of −1.66 V), and silver will be reduced (with standard potential of +0.8 V). Note that the solution used in the silverware experiment is an alkaline solution. It serves two functions. First, it allows the passage of ionic currents. Second, it dissolves the aluminum hydroxide (an electrical insulator) formed on the aluminum surface during the electrochemical reaction. In this way, a continuous electrical circuit is maintained throughout the electrochemical reaction.

8.4 INFLUENCE OF CONCENTRATION AND TEMPERATURE ON ELECTRODE POTENTIALS

For conditions other than standard, we can calculate the electrode potential relative to the standard potential using the Nernst equation:

$$V = V_o + \frac{RT}{nF} \ln C, \tag{8.2a}$$

where

V = the electrode potential
V_o = standard electrode potential
R = the gas constant (8.3 J/mol-K)
F = the Faraday constant (eN_A = 96,500 Coul./mol; N_A = Avogadro's number)
T = the absolute temperature
C = the molar concentration of ions
n = the number of electrons involved in the half-cell reaction

EXAMPLE

For the Zn/Zn^{2+} half-cell reaction, calculate the electrode potential at 250 K at Zn^{2+} concentration of 0.1 M.

Solution

For the Zn/Zn^{2+} half-cell reaction, the standard potential V_o = –0.76 V and n = 2 since two electrons are removed from Zn to form Zn^{2+}. From Equation (8.2a), we can write:

$$V = -0.76 + \frac{8.3 \times 250}{2 \times 96500} \ln(0.1)$$

$$= -0.76 - 0.025$$

$$= -0.785 \text{ V.}$$

At room temperature (298 K), this equation can be simplified to:

$$V = V_o + \frac{0.059}{n} \log_{10} C. \tag{8.2b}$$

Note that the electrode potential becomes more negative at lower ion concentration. This indicates that the electrode in contact with a solution containing lower ion concentration is more likely to corrode. For example, consider a piece of zinc immersed in HCl. Because of random fluctuations in concentration or liquid access to the zinc surface, some parts of zinc are in contact with the HCl solution having higher concentrations of Zn^{2+} and some with those containing lower concentrations. The Nernst equation shows that zinc dissolution is most likely where the Zn^{2+} concentration is the lowest. We will return to this point in later discussions on conditions for corrosion.

Question for discussion: Normally, when we drop a piece of zinc into HCl, zinc dissolves with the evolution of hydrogen. Can you think of ways to make the reaction go in the opposite direction — that is, zinc chloride reacting with hydrogen to deposit zinc?

8.5 POWER BY CORROSION: THE Cu–Zn BATTERY

Many batteries operate on a principle similar to the Cu–Zn battery, as illustrated in Figure 8.4, except that the hydrogen half-cell is replaced by a copper electrode in contact with a solution containing copper ions. The two half-cell reactions are:

$$\text{Zn} \rightarrow \text{Zn}^{2+} + 2e^- \ (V_o = -0.76\text{V}); \tag{8.3a}$$

$$\text{Cu}^{2+} + 2e^- \rightarrow \text{Cu} \ (V_o = -0.34\text{V}). \tag{8.3b}$$

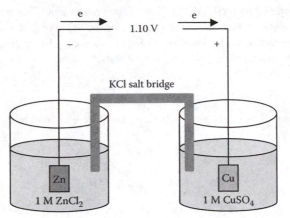

FIGURE 8.4 The Cu–Zn battery based on Zn in contact with 1 M $ZnCl_2$ solution and Cu in contact with 1 M $CuSO_4$ solution at 25°C.

Note the sign reversal for Equation (8.3b) when we invert the equation from Table 8.1. In a standard configuration, Zn has a stronger tendency to give off electrons than Cu. Zn is the species to be oxidized. Adding Equation (8.3a) and Equation (8.3b), we obtain a standard electrode potential of 1.10 V (negative on Zn), and the net reaction is:

$$Zn + Cu^{2+} \rightarrow Zn^{2+} + Cu. \tag{8.4}$$

During battery operation, the zinc electrode produces Zn^{2+} ions as indicated in Equation (8.3a) and Equation (8.4), which go into solution, while Cu^{2+} ions from solution form metallic copper on the copper electrode, according to Equation (8.3b) and Equation (8.4). This is known as electrodeposition or electroplating.*

8.5.1 ENERGY AND VOLTAGE

Associated with a given standard half-cell reaction, there is a free energy change and a standard electrode potential. The relationship is given by:

$$\Delta G = neV_o N_A$$
$$= nFV_o, \tag{8.5}$$

where
     ΔG = free energy change associated with the reaction
     V_o = standard electrode potential
     n = number of electrons associated with the reaction
     e = electron charge (1.6×10^{-19} C)
     N_A = Avogadro's number (6.02×10^{23})
     F = Faraday constant (eN_A = 96,500 Coul.).

---

* There is an old story about the black magic of chromium electroplating, which produces a shiny coating against oxidation. A foreman in a chromium-plating company retired after many years of service. After his retirement, the quality of surface finish of the chromium plating was never the same. The new foreman was puzzled, since nothing was changed — the same plating bath, same chemicals, same procedures, etc. He consulted with the retired foreman, explored every detail of his workday, and discovered, to the new foreman's horror and delight, an unsanitary habit of the old man at the beginning of each workday after several cups of coffee. After several trial-and-error runs, the new foreman worked out an optimum concentration of urea as one component of the chromium-plating bath, and the superior surface finish returned.

This relationship can be understood as follows. The energy available from the reaction to move n electrons up the potential hill is ΔG per molecule, or $\Delta G/N_A$. For potential V_o, the height of the potential hill is eV_o, where e is the electron charge. Therefore, for n electrons, the total energy needed to move up the potential hill is neV_o, which must be equal to $\Delta G/N_A$.

EXAMPLE

Consider a battery made of two half-cells involving Ag and Al (our silverware experiment). The relevant half-cell reactions are:

$$Al \rightarrow Al^{3+} + 3e \ (V_o = -1.66 \text{ V});$$

$$Ag \rightarrow Ag^+ + e \ (V_o = +0.80 \text{ V}).$$

What is the standard voltage from a silver–aluminum battery resulting from the following electrochemical reaction?

$$Al + 3Ag^{3+} \rightarrow Al^{3+} + 3Ag.$$

Solution

Let us rewrite the half-cell reactions as follows:

$$Al \rightarrow Al^{3+} + 3e \ (V_o = -1.66 \text{ V}); \tag{a}$$

$$Ag^+ + e \rightarrow Ag \ (V_o = -0.80 \text{ V}). \tag{b}$$

Now multiply the Ag half-cell reaction by 3:

$$3Ag^+ + 3e \rightarrow 3Ag. \tag{c}$$

What is the standard electrode potential for this reaction? Is it –0.80 V or $-0.80 \times 3 = -2.40$ V? The answer is –0.80 V. Recall Equation (8.5). The standard electrode potential already takes into account the number of electrons. There is no need to multiply the electrode potential. Adding Equations (a) and (c) and canceling three electrons on both sides, we have:

$$Al + 3Ag^{3+} \rightarrow Al^{3+} + 3Ag.$$

The standard electrode potential for this reaction is $-1.66 - 0.8 = -2.46$ V. The negative sign indicates that Al is the negative electrode.

Question for discussion: Consider a battery made of the configuration: Zn|ZnCl$_2$ solution|Cu. On the zinc electrode, we have the same half-cell reaction as shown in Equation (8.3a). On the copper side, the half-cell reaction shown in Equation (8.3b) cannot occur because no copper ions are available to receive the electrons. We need another reduction reaction to consume the electrons. There are at least three (Equations 8.6a, b, c).

$$H^+ + e^- \rightarrow 1/2 \ H_2 \ (V_o = 0.00 \text{ V}); \tag{8.6a}$$

$$O_2 + 2H_2O + 4 \ e^- \rightarrow 4OH^- \ (V_o = -0.40 \text{ V}); \tag{8.6b}$$

$$O_2 + 4H^+ + 4 \ e^- \rightarrow 2H_2O \ (V_o = -1.23 \text{ V}). \tag{8.6c}$$

Equations (8.6a) and (8.6c) are favored in an acidic solution (high concentration of H^+ or low pH), while Equations (8.6b) and (8.6c) require the presence of dissolved oxygen. Discuss how the voltage output under standard conditions can range from 0.76 to 1.99 V, depending on the pH and the amount of dissolved oxygen in the ZnCl$_2$ solution.

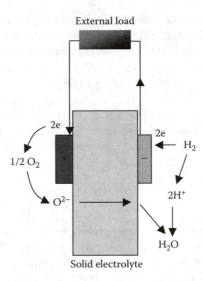

FIGURE 8.5 The hydrogen fuel cell.

8.6 THE HYDROGEN FUEL CELL

In a hydrogen fuel cell (Figure 8.5), hydrogen (the fuel) is exposed to a platinum electrode, dissociating to produce H^+ and electrons. The hydrogen side becomes the negative terminal of the fuel cell. On the other side of the cell, each oxygen atom adsorbs two electrons from the electrode surface and becomes an oxygen anion (O^{2-}). This electrode is therefore the positive terminal of the fuel cell. The oxygen anion diffuses through an ionic conductor to the hydrogen side, combining with hydrogen atoms to form water and releasing two electrons. Oxides such as alumina and zirconia are used as oxygen ionic conductors. The electrical current drawn from a fuel cell is limited by the speed of ionic conduction, which is controlled by diffusion. As a result, fuel cells are normally operated at elevated temperatures (as high as 800°C) to increase diffusion rates. The two half-cell reactions are:

$$1/2\ O_2 + 2e^- \rightarrow O^{2-}; \tag{8.7a}$$

$$O^{2-} + 2H \rightarrow H_2O + 2e^-. \tag{8.7b}$$

The net reaction is the production of water from hydrogen and oxygen*:

$$1/2\ O_2 + H_2 \rightarrow H_2O. \tag{8.8}$$

The free energy release is 237 kJ/mol. From Equation (8.5), we can write:

$$237 \times 10^3 = 2 \times 96,500 \times V_o,$$

where V_o is the standard cell voltage. Solving, we find that $V_o = 1.23$ V.

---

* Here is an easy way to demonstrate the operation of a fuel cell. Set up a normal electrolysis cell (e.g., sodium hydroxide solution with two platinum electrodes). Apply a few volts between the two platinum electrodes to electrolyze the water, with hydrogen bubbling from the negative terminal and oxygen from the positive terminal. After a minute or so, stop. Gas bubbles still cling to the electrodes. Now attach a voltmeter across the two electrodes. You will see a voltage that appears because of the reverse reaction, equation (8.7). Eventually, the voltage will decay to zero as the "fuel" is being consumed.

In terms of efficiency and cleanliness, the hydrogen fuel cell is probably one of the most desirable energy sources. The major problem is the cost and availability of hydrogen. In principle, hydrogen can be obtained from hydrocarbons such as methane as follows:

$$CH_4 + H_2O \rightarrow CO + 3H_2. \tag{8.9}$$

This is known as a steam-reforming reaction and is thermodynamically favorable at elevated temperatures (i.e., heat from another energy source is needed to sustain this reaction). A better solution is to use a hydrocarbon such as methane or methanol directly as fuel input for the fuel cell. This continues to be an active area of research.

8.7 RUSTING OF IRON

Have you ever wondered why rust occurs more readily in humid climates? Even more intriguing, have you noticed that rust does not occur uniformly on a piece of iron, but tends to form near cracks and crevices? To answer these questions, note that three components are required for the rusting of iron: iron, water, and oxygen. The two half-cell reactions to consider are:

$$Fe \rightarrow Fe^{2+} + 2e^- \ (V_o = -0.44 \text{ V}); \tag{8.10a}$$

$$O_2 + 2H_2O + 4e^- \rightarrow 4OH^- \ (V_o = -0.40 \text{ V}). \tag{8.10b}$$

Figure 8.6 shows the overall corrosion process. Fe turns into Fe^{2+}, while oxygen and water react to form OH^-, followed by the reaction between Fe^{2+} and OH^- to form ferrous hydroxide. The standard electrode potential is 0.84 V. The initial corrosion product is ferrous hydroxide, which is subsequently oxidized to form hydrated ferric hydroxide (rust).*

How do we explain the nonuniformity of rust formation? When all parts of the iron surface are equally accessible to water and oxygen, the iron surface forms rust uniformly. In the real world, the ideal condition of uniform oxygen exposure may not occur. Consider Equation (8.10b), the second half-cell reaction. This reaction consumes electrons. The higher the oxygen concentration is, the more electrons it will consume. Therefore, this reaction takes electrons from regions with lower oxygen concentration. The lower oxygen concentration region supplies electrons via the transformation of Fe to Fe^{2+}. Therefore, corrosion occurs preferentially in places where the oxygen concentration is low or oxygen access is poor. These are usually cracks, crevices, or locations covered by dirt, grease, etc. The term "crevice corrosion" is often used to describe corrosion occurring in places that are hard to reach.

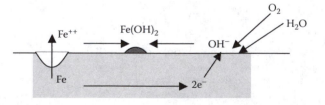

FIGURE 8.6 Rusting of iron.

---

* Heat is released during corrosion reactions. This phenomenon is utilized in commercial heat packs. Each heat pack consists of fine iron powders, water, and salt, packed under nitrogen. Upon opening the pack, air diffuses into the pack, dissolves in water, triggers the formation of rust, and hence releases heat.

8.8 CONDITIONS FOR CORROSION

Corrosion occurs when there are regions of heterogeneities (presence of anode and cathode) through which electrical currents pass. There are three common instances for such heterogeneities.

8.8.1 COMPOSITION DIFFERENCE

When two dissimilar materials are present and in electrical contact, corrosion may occur since the electrode potentials for these materials are most likely different. The material with the more negative electrode potential is the anode (corroding electrode).

For example, Pb–Sn solder is used to connect Cu wires. Given that the electrode potentials for these three materials are different (Pb: –0.13 V; Sn: –0.14 V; Cu: +0.34 V), corrosion of the solder material is expected to occur in humid air. Tin cans use tin plating to protect the underlying iron surface from corrosion. However, as soon as the tin plating is scratched, the iron surface is exposed to the environment, and corrosion of the underlying iron surface occurs (the standard electrode potential is –0.44 V for Fe and –0.14 V for Sn). In an aluminum–copper alloy with $CuAl_2$ precipitates (the θ-phase), the aluminum matrix has a more negative electrode potential than the precipitates. This potential difference drives the corrosion of the matrix.

8.8.2 STRESS

Heterogeneities can also be induced by mechanical deformation or changes in nearest-neighbor atomic environments. When a metal is under tensile or compressive stress, the free energy of the stressed region increases. As a result, the free energy change ΔG from the initial state (metal atoms under mechanical stress) to the final state (metal ions) is larger. According to Equation (8.5), the electrode potential becomes more negative. This means that the stressed material becomes the anode when exposed to an environment conducive to corrosion. This is another good example of the relationship between mechanics and chemistry, as discussed in the chapter on mechanical properties.

For example, when we immerse a polycrystalline sample in an etching solution, grain boundaries are etched preferentially. Similarly, dislocations show up as etch pits. We may consider grain boundaries and dislocations as microscopically stressed regions. Another example can be demonstrated by dipping a bent metal wire in a corrosive environment (e.g., salt or acid solution) (Figure 8.7). The bent portion of the wire is preferentially attacked. Stress-induced corrosion in metallic prosthetic implants (e.g., hip and knee joints) is a concern due to the large stresses imposed on these components.

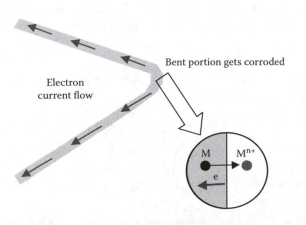

FIGURE 8.7 Corrosion of a bent metal wire. M = metal atom; M^{n+} = metal ion; e = electron.

A Closer Look

Consider a piece of iron being deformed with an elastic strain of 2%. The elastic modulus of iron is 200 GPa. What is the change in the standard electrode potential for the Fe/Fe^{2+} half-cell reaction under this condition? This can be determined by first calculating the elastic strain energy per unit volume U as follows:

$$U = \frac{1}{2}E\varepsilon^2,$$

where E is the elastic modulus and ε the elastic strain. This can be evaluated to give:

$$U = \frac{1}{2}200 \times 10^9 \times (0.02)^2$$

$$= 4 \times 10^7 \ J/m^3$$

$$= 40 \ J/cm^3.$$

We need to convert the energy per unit volume into energy per mole u as follows:

$$u = U\frac{A}{\rho},$$

where A is the atomic weight and ρ the mass density. For iron, $A = 56$ and $\rho = 8$ g/cm^3. Substituting, we find that:

$$u = U\frac{A}{\rho} = 40 \times \frac{56}{8}$$

$$= 280 \ J/mole.$$

Because of this strain energy contribution, the free energy change in the Fe/Fe^{2+} half-cell reaction becomes more negative. The change in electron potential ΔV is given by Equation (8.5):

$$\Delta V = -\frac{u}{nF}$$

$$= -\frac{280}{2 \times 96500}$$

$$= -0.00145 \ V.$$

Remember: A stressed solid has a more anodic (negative in our definition) potential than the unstressed solid.

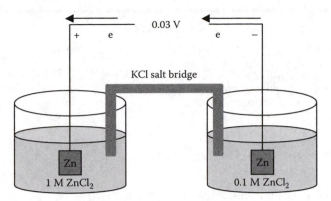

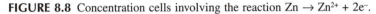

FIGURE 8.8 Concentration cells involving the reaction $Zn \rightarrow Zn^{2+} + 2e^-$.

8.8.3 CONCENTRATION DIFFERENCE

The Nernst equation shows that the electrode potential decreases with a decreasing concentration of the species required for corrosion. For example, the standard electrode potential for the half-cell reaction $Zn \rightarrow Zn^{2+} + 2e^-$ is -0.76 V (Zn in contact with 1 M $ZnCl_2$ solution at room temperature). If the concentration of the $ZnCl_2$ solution is 0.1 M, the electrode potential for the same reaction falls to -0.79 V (Figure 8.8). The half-cell with lower Zn^{2+} concentration becomes the anode, and the corresponding Zn electrode dissolves. One way to think about this is that the system wants to attain the same Zn^{2+} concentration in the two concentration half-cells. This is accomplished by the dissolution of the Zn electrode in the lower concentration $ZnCl_2$ solution and the deposition of Zn in the higher concentration $ZnCl_2$ solution.

Concentration difference plays a central role during the rusting of iron in cracks and crevices, as discussed earlier. These cracks and crevices restrict oxygen access and create oxygen concentration gradients. This oxygen concentration difference creates electrode potential differences that determine the (corroding) anode location.

8.9 RATE OF CORROSION

The rate of corrosion can be obtained from two pieces of information: the equation describing the corrosion reaction and the corrosion current. The thinning rate of the corroded metal electrode TR is given by:

$$TR = \frac{I}{n} \frac{A}{Fa\rho},$$

(8.11)

where

      I = the corrosion current (amp)
      n = number of electrons involved in the electrochemical reaction
      A = the atomic weight
      F = the Faraday constant (96,500 Coul.)
      a = the exposed area of the metal electrode
      ρ = its density

Note that when the anode has a small area a (such as the case of tiny cracks and crevices), the thinning rate becomes large, leading to the formation of a pit.

EXAMPLE

Consider the $Zn|ZnCl_2$–$Cu|CuSO_4$ reaction, as shown in Figure 8.4; Zn is corroded in this reaction. The corrosion current is 0.5 amp and $n = 2$. The atomic weight of Zn is 65.4 and its mass density is 7.14 g/cm^3. Calculate the thinning rate of an exposed zinc electrode area of 1 cm. If the thickness of the zinc electrode is 0.5 cm in this Zn–Cu battery, what is the battery's lifetime?

Solution

Substitute into Equation (8.11) the following quantities:

$I = 0.5$ amp
$n = 2$
$A = 65.4$
$F = 96,500$
$a = 1$ cm
$\rho = 7.14$ g/cm^3

We have:

$$TR = \frac{0.5}{2}\frac{65.4}{96500 \times 1.0 \times 7.14}$$

$$= 2.37 \times 10^{-5} \text{ cm/s.}$$

At this thinning rate, the lifetime is $0.5/(2.37 \times 10^{-5}) = 21071$ s, or a little less than 6 h.

A Closer Look

The thinning rate is determined by the corrosion current, but what determines the corrosion current? In a normal electrical circuit, electrical current depends on the applied voltage and the electrical resistance of the circuit. In a corrosion situation, the corrosion current at a given overvoltage (i.e., voltage over and above the standard potential*) is controlled by two factors: activation barrier and concentration depletion.

Similar to other chemical reactions, the corrosion rate is limited by some sort of an activation barrier. This phenomenon is known as activation polarization. For example, in the H_2/H^+ half-cell biased at some potential, H_2 forms and this results in the evolution of hydrogen gas. This reaction may have to overcome an activation barrier due to surface diffusion of atomic hydrogen or formation of a gaseous layer that blocks migration of H^+ to the electrode surface. The relationship between overvoltage (i.e., voltage above the standard electrode potential) due to the activation barrier ΔV and the corrosion current density J is given by:

$$\Delta V = V - V_o = \beta \cdot \log_{10}\left(\frac{J}{J_o}\right), \tag{8.12}$$

where J_o and β are constants for a given half-cell. J_o is known as the exchange current density, V the half-cell potential, and V_o the standard potential; β is a negative number for reduction reactions and a positive number for oxidation reactions.

---

* At the standard potential under standard conditions, the system is at equilibrium, and there is no current flow.

For corrosion reactions requiring the arrival of ions from the solution to the electrode surface (as in reduction reactions, such as copper plating: $Cu^{2+} + 2e \rightarrow Cu$), the reaction rate is limited by the available concentration of ions in solution. This may be due to low initial ion concentration or a depletion of ions near the electrode (agitation of the solution should minimize this effect). In this case, the corrosion current saturates beyond a certain overvoltage. This phenomenon is known as concentration polarization.

Corrosion of metals and alloys is often studied by measuring how the corrosion current varies with voltage applied to the material of interest. The resulting trace as shown schematically in Figure 8.9 is known as a potentiodynamic scan. Note that V_t marks the transition between reduction and oxidation.*

8.10 CORROSION CONTROL

There are primarily two methods for corrosion control (Figure 8.10).

Passivation. Corrosion can be eliminated by isolating the active surface from the environment or eliminating the electrical path needed for corrosion currents to flow. This can be accomplished by covering the active surface with an adherent, usually an insulating and impermeable coating (e.g., oxide, paint, etc.). Alternatively, a perfectly dry environment can be maintained to eliminate the availability of water vapor. In some cases, the corrosion product provides the natural passivation layer. For example, aluminum oxide is formed on the surface of aluminum as a result of oxidation or corrosion.** Such an oxide layer tends to be adherent and continuous and is an electrical insulator. This stops the flow of corrosion currents and hence inhibits corrosion.***

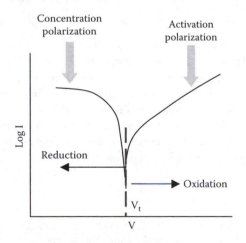

FIGURE 8.9 A typical potentiodynamic scan, showing the occurrence of oxidation and reduction.

---

* In case you are wondering how to deduce the corrosion current, try the problem section.

** Al_2O_3 has a hexagonal face with the Al–Al spacing equal to 0.267 nm. This is almost identical to the Al–Al spacing on the (111) plane of Al (0.286 nm). This coherent arrangement of atoms creates a nearly stress-free and strong interface between Al and Al_2O_3.

*** To improve the corrosion protection properties of aluminum oxide, the thickness can be increased by a process called anodization. Place the aluminum sample in an electrolyte (typically 1–3 M sulphuric acid plus 1–2 vol% of oxalic acid) and apply a positive voltage V to the aluminum sample (negative voltage to a counter electrode). An aluminum oxide layer is formed on the aluminum sample, with the thickness directly related to the voltage V. By adjusting the current density and temperature during anodization, the size and concentration of pores in this oxide layer can be controlled. Subsequent dipping of the anodized specimen into an appropriate dye solution can produce any color finish desired.

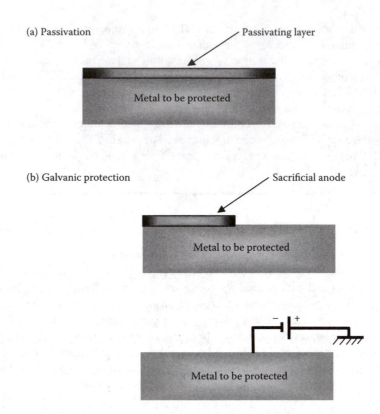

FIGURE 8.10 Methods of corrosion control: passivation and galvanic protection.

Galvanic protection. Here, the idea is to force the material to be protected to become the cathode. This can be achieved in two ways. One is to use a sacrificial coating or material in electrical contact with the surface to be protected — for example, zinc coating in galvanized steel or installing a magnesium anode in a water tank made of steel. In both cases, the electrode potential of the sacrificial material is more negative than that of the material we wish to protect and is corroded instead. The other method is to apply a negative voltage to the material to be protected, forcing the material to become the cathode (cathodic protection).

8.11 OXIDATION

For most metals at room temperature, the free energy for the formation of the metal oxide is negative, meaning that such oxidation reactions are favorable. In some cases, these oxides form passivating layers protecting the underlying metals (e.g., aluminum oxide); in other cases, they do not (e.g., iron oxides). In this section, we explore the process of oxidation and the conditions under which oxides may form protecting layers against corrosion.

When we measure the weight gain Δm of a metal due to oxidation as a function of time t, the results generally fall into two categories: (1) $\Delta m \propto t^{1/2}$; and (2) $\Delta m \propto t$. In the first category (parabolic growth), the square root dependence suggests a diffusion-based mechanism. There are two possibilities, depending on the relative diffusion rates of the metallic and oxygen ions, as shown in Figure 8.11.

In the mechanism shown in Figure 8.11(a), oxygen ions diffuse through the oxide and react with the metal to produce a thicker oxide layer. The oxide grows at the metal–oxide interface (e.g., Ti and Zr). In the mechanism shown in Figure 8.11(b), metal ions diffuse through the oxide layer

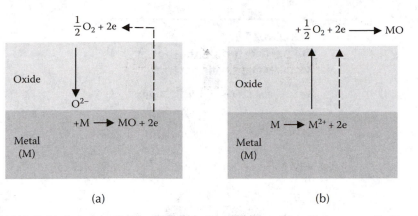

(a)   (b)

FIGURE 8.11 (a) Oxidation controlled by diffusion of oxygen ions; (b) oxidation controlled by diffusion of metal ions.

and react with oxygen ions at the oxide–oxygen interface to produce a thicker oxide layer. The oxide grows at the oxide–oxygen interface (e.g., Fe, Cr, and Cu). Note that the second mechanism results in the production of vacancies in the metal substrate.

The linear growth mechanism can be rationalized by assuming that the oxide film develops cracks during growth, as a result of excessive growth stress. Cracks in the oxide allow direct access of oxygen to the metal. Oxidation can then proceed without requiring the diffusion of metal or oxygen ions. The amount of oxide formed then depends linearly on the time of oxygen exposure.

During oxide growth, when the volume of the oxide per metal atom is greater than that of the metal, the resulting stress is compressive and may cause parts of the film to delaminate. Conversely, when the volume of the oxide per metal atom is less than that of the metal, the resulting stress is tensile and may cause cracking. The ideal situation to prevent delamination or cracking is when the molar volume of the metal is equal to the molar volume of the metal oxide per metal atom. For the oxidation reaction $xM + yO_2 \rightarrow M_xO_{2y}$ involving metal M, 1 mol of M produces $(1/x)$ mol of the oxide M_xO_{2y}. The ratio of the volume of the product (metal oxide) per metal atom to that of the reactant (metal) is known as the Pilling–Bedworth ratio Ω, as given by:

$$\Omega = \frac{\left(\dfrac{A_o}{x\rho_o}\right)}{\left(\dfrac{A_m}{\rho_m}\right)}, \tag{8.13}$$

where

      A is the atomic weight
      ρ is the density
      subscripts o and m represent the oxide and metal, respectively
      x is the number of metal atoms in the metal oxide formula

Metal/metal oxide systems with Ω close to 1 should form protective oxides, while those with Ω significantly different from 1 probably would not. Table 8.2 lists the Pilling–Bedworth ratios for several selected systems. A Pilling–Bedworth ratio near 1 does not always guarantee a protective oxide in actual applications. Other factors, such as adhesion and thermal expansion match between the metal and the metal oxide, abrasion/erosion resistance of the oxide, and its chemical stability, are sometimes more important.

TABLE 8.2
Pilling–Bedworth Ratios for Several Metal/Metal Oxide Systems

Metal/Metal Oxide	A_o/ρ_o	A_m/ρ_m	Ω
Al/Al_2O_3	25.6	10	1.28
Ti/TiO_2	18.9	10.6	1.78
Cr/Cr_2O_3	29.2	7.3	2.00
Fe/Fe_2O_3[a]	30.5	7.1	2.15
Ta/Ta_2O_5[a]	53.9	10.9	2.47

[a] Nonprotective.

8.12 A FEW EXAMPLES FOR THOUGHT

8.12.1 BATTERIES FOR ELECTRIC VEHICLES: ENERGY CAPACITY ANALYSIS

In this section, we will attempt to compare the energy capacity of gasoline with batteries and therefore determine to what extent electric vehicles are viable. In this analysis, it is convenient to express energy in units of kilowatt-hours (1 kWh = 3.6×10^6 J).

The raw energy content of gasoline is about 13 kWh/kg or ~10 kWh/L. Under average driving conditions, the overall conversion efficiency of chemical energy in gasoline to mechanical energy with current internal combustion engines is about 20%. Therefore, the effective energy content of gasoline is 2.6 kWh/kg, or ~8.0 kWh/U.S. gallon.* Standard Pb acid batteries have a usable energy capacity of 35 Wh/kg. This is a factor of 75 less than that for gasoline.

Given a car weighing 1000 kg with an average drag or friction coefficient of 0.1 (i.e., the drag force ~1000 N), we can show readily that the energy consumed is about 0.4 kWh/mile (corresponding to 20 miles per gallon of gasoline). If this car is powered by batteries weighing 400 kg (900 lb) with an energy density of E Wh/kg, the total energy content = 0.4E kWh. The range of this electric vehicle is 0.4E/0.4 = E miles. Therefore, a good rule of thumb is that for every 1 Wh/kg energy capacity, a range of 1 mile is obtained. Therefore, we expect to achieve ranges ~35 miles using Pb acid batteries.

For practical transportation purposes, we need to have ranges ~200 miles or greater. Current Ni–Cd batteries have energy capacities ~100 Wh/kg, and Li ion batteries have energy capacities ~200 Wh/kg. Therefore, purely from the energy capacity point of view, Li ion batteries are the only viable conventional batteries today for electric vehicles. Of course, other factors need to be considered as well: number of charge–discharge cycles the battery can endure, cost, recycling, and disposal.

---

* One U.S. gallon = 3.785 L; specific gravity of gasoline = 0.8. Using the combustion of octane as an example, we can write: C_8H_{18} + (25/2) $O_2 \rightarrow 8CO_2$ + $9H_2O$. From the values of standard enthalpy of formation (–242 kJ/mol for water vapor, –394 kJ/mol for CO_2, and –250 kJ/mol for liquid octane), the standard enthalpy for the combustion reaction is –5080 kJ/mol of liquid octane or –44.6 kJ/g. One liter of fuel weighs 800 g, and its energy content is therefore 44.6×800 = 35.6 MJ or about 10 kWh. This also means that the ideal oxygen-to-fuel weight ratio is 400:114, or 3.5. Air contains 23% of oxygen by weight at sea level in a standard atmosphere (temperature = 15°C and pressure = 1013.2 Pa). Therefore, the ideal air-to-fuel weight ratio is 3.5:0.23, or about 15. Since air density drops about 3% for every 1000-ft increase in elevation, a perceptible decrease in car performance will be noted during a drive from Chicago (650 ft above mean sea level) to Denver (5000 ft above MSL). For most aircraft engines, the fuel mixture is typically leaned when operating above 3000–5000 ft MSL in order to obtain the best fuel economy.

8.12.2   CARBON FUEL CELLS?

In the preceding section, we discussed the use of hydrogen fuel cells as an alternative source of energy. While combustion of hydrogen is clean, the process to obtain hydrogen may not be (unless wind or solar energy is used to produce the hydrogen fuel). In principle, any other combustion process can be utilized to produce electrical energy — for example, the combustion of carbon to produce CO_2. In this process, the free energy release under standard conditions is 394 kJ/mol. Since four electrons are involved in this oxidation process, it can be shown that such a carbon fuel cell develops a standard voltage of 1.03 V.

If developed with efficiency similar to the hydrogen fuel cell, such a carbon fuel cell will have several advantages. First, it produces less CO_2 per unit energy generated than conventional power plants. Second, solid carbon is readily portable. Third, the reaction product is gaseous, thus requiring no disposal. Fourth, the energy density is high. A laptop computer under typical use consumes about 100 kJ/h, or about 400 kJ in a 4-h plane trip. If this laptop computer were powered by a carbon fuel cell at 50% energy conversion efficiency, only 25 g of carbon would be consumed.

8.12.3   CORROSION CONCERNS FOR PROSTHETIC IMPLANTS

Since body fluid is an aqueous electrolyte, we have to worry about corrosion of any metallic implants, especially stress-induced corrosion. If oxides are part of the corrosion products, the implant materials should have oxides that are adherent and continuous. This is particularly difficult when the material is subjected to high shear stress in use that may shear off the oxide. In any case, if the corrosion current density J is known, it is possible to deduce the thinning rate — hence, the lifetime of an implant — using Equation (8.11). As an example, let us assume that the corrosion current density $J = 100$ $\mu A/cm^2$, the number of electrons involved in the corrosion reaction $n = 4$; the atomic weight of the metal $A = 45$; and the metal density $\rho = 4.5$ g/cm^3. Equation (8.11) shows that this metal is being corroded at the rate of 0.08 cm per year. At this point, how much loss is tolerable before the implant loses its functionality must be determined.

8.12.4   CORROSION PROTECTION IN HARD-DISK DRIVES

Information in a hard-disk drive is contained in a magnetic cobalt alloy thin film about 25 nm thick (in 2002). A carbon-based overcoat is deposited on top of the magnetic layer and is about 3–4 nm thick (in 2002). This overcoat provides wear and corrosion protection for the underlying magnetic layer. If the corrosion loss of 10 nm of the cobalt layer in 5 years can be tolerated, we require the corrosion current density to be approximately 1/4 nA/cm^2, an incredibly small value. The protective overcoat must therefore be synthesized to be dense, uniform, and free of pinholes at the atomic scale, a feat that is repeated millions of times a day in the hard-drive industry.

8.12.5   PROPULSION BY OXIDATION

When oxidized, 1 g of aluminum produces a free energy of ~7.5 kcal, or 31 kJ. Normally, aluminum does not oxidize rapidly in air because of the low oxygen concentration and slow ionic diffusion through the oxide. We can accelerate this process by mixing fine aluminum powders (to provide more surface area for faster reaction and minimize diffusion limitation) with solid oxygen (in the form of nitrates or perchlorates). Indeed, that is the fuel used in the space shuttle's solid rocket booster (mixture of aluminum powders and ammonium perchlorate with some iron oxides as catalyst).

Assume that we can combust 1 kg of Al powder per second. This results in a final mass of 1.89 kg of Al_2O_3 and a free energy release of ~3×10^7 J/s. If 100% of the energy is used to expel the aluminum oxide particles, the velocity of the alumina particles will be about 5.8×10^3 m/s, resulting in a thrust of ~1.1×10^4 N.* This method of propulsion is not very efficient compared with that of commercial jets. For example, to obtain the same combined thrust of 4.4×10^5 N produced by the two turbofan engines that power a typical Boeing 767, it is necessary to combust 40 kg of Al per second. This is about 40 times the fuel burn rate of the 767.

8.13 COMMON BATTERIES

8.13.1 LEAD–ACID

The lead–acid battery consists of lead (Pb) and lead dioxide (PbO_2) electrodes immersed in sulfuric acid. The two half-cell reactions are:

$$Pb + HSO_4^- \rightarrow PbSO_4 + H^+ + 2e; \tag{8.14a}$$

$$PbO_2 + HSO_4^- + 3H^+ + e \rightarrow PbSO_4 + 2H_2O. \tag{8.14b}$$

The net result is the conversion of lead and lead dioxide into lead sulfate. The open-circuit voltage for this reaction under standard conditions is slightly above 2 V. A modern lead–acid battery is reliable, exhibits no memory effects, requires almost no maintenance, and is readily recyclable. The major drawback is its relatively low energy capacity (~35 Wh/kg), which is due more to the lack of research and development than any inherent limitation.

8.13.2 ALKALINE

The modern alkaline battery is an improvement of the old carbon–zinc "dry cell." The positive terminal of a carbon–zinc battery is MnO_2 impregnated into a conducting graphite rod. The negative terminal is zinc. The electrolyte is a gel mixture of zinc chloride and ammonium chloride. The relevant half-cell reactions are:

$$Zn \rightarrow Zn^{2+} + 2e; \tag{8.15a}$$

$$2MnO_2 + 2NH_4^+ + 2e \rightarrow Mn_2O_3 + H_2O + 2NH_3. \tag{8.15b}$$

The open-circuit voltage for this battery is about 1.5 V. Notice that ammonia gas is being produced. In addition, another side reaction is producing gaseous ammonia and hydrogen:

$$2NH_4^+ + 2e \rightarrow 2NH_3 + H_2. \tag{8.15c}$$

Because of the buildup of gas pressure, the battery eventually leaks, creating a rather messy situation.** This problem is solved by replacing the ammonium chloride with potassium hydroxide (hence the name "alkaline battery"). In this case, the appropriate half-cell reactions are:

$$Zn + 2OH^- \rightarrow Zn(OH)_2 + 2e; \tag{8.16a}$$

$$2MnO_2 + H_2O + 2e \rightarrow Mn_2O_3 + 2OH^-. \tag{8.16b}$$

As can be seen, no gas is produced in an alkaline cell.

---

* Thrust is equal to the negative rate of change of linear momentum of the ejected aluminum oxide powders.
** Reactions involving $ZnCl_2$ and MnO_2 can help to remove ammonia and hydrogen. Unfortunately, they are not sufficient to remove these gases completely.

8.13.3 Ni–Cd

As the name implies, the battery has Ni and Cd as the electrode materials, with potassium hydroxide as the electrolyte. The half-cell reactions are:

$$Cd + 2OH^- \rightarrow Cd(OH)_2 + 2e; \tag{8.17a}$$

$$NiOOH + H_2O + e \rightarrow Ni(OH)_2 + OH^-. \tag{8.17b}$$

Nickel oxyhydroxide (NiOOH) is formed in the presence of oxygen and an alkaline solution. The operating voltage is typically around 1.25 V. If used properly, Ni–Cd batteries can withstand many charge–discharge cycles. The major disadvantage is the memory effect: If charged before the cell is fully discharged, the cell tends to develop lower energy storage capacity.

8.13.4 Ni–MH (Metal Hydride)

Ni–MH is an advance over Ni–Cd in two ways: reduced memory effects and about 40% higher energy capacity (~100 Wh/kg). The electrochemistry is similar to that for Ni–Cd, except that a metallic alloy is used in place of Cd. Examples of such metallic alloys include $LaNi_5$ and $TiZr_2$ that absorb hydrogen to form a metal hydride readily. In fact, such alloys are used for hydrogen storage applications. The electrolyte is potassium hydroxide. The appropriate half-cell reactions are:

$$M–H + OH^- \rightarrow M + H_2O + e; \tag{8.18a}$$

$$NiOOH + H_2O + e \rightarrow Ni(OH)_2 + OH^-. \tag{8.18b}$$

The operating voltage is about 1.25 V. Note that the metal hydride is formed during charging.

8.13.5 Lithium Ion

The lithium-ion battery produces the highest operating voltage of all commercially available batteries, about 4.0 V. The negative terminal of a typical lithium-ion battery consists of $LiCoO_2$, and the positive terminal is graphite. During operation, lithium ions (Li^+) are transported from $LiCoO_2$ through an electrolyte to graphite. Because of its open structure, graphite can store large amounts of lithium without falling apart. The lithium-ion battery provides the highest energy density (~200 Wh/kg) of all commercially available batteries, has low self-discharge rates, and does not exhibit memory effects.

PROBLEMS

1. You are given the following standard electrode potential information:

Element	Standard potential (volt)
Cu	+0.34
Pb	−0.13
Sn	−0.14

    a. Sn/Pb solder is used to connect Cu wires on circuit boards. Describe the potential corrosion problem in such a system in contact with moist air.

    b. Suggest one method to solve this problem (still using exactly the same materials).

2. Consider a Ag–Zn battery. The two relevant half-cell reactions are:

$Ag \rightarrow Ag^+ + e$ $(V_o = +0.80$ V)

$Zn \rightarrow Zn^{++} + 2e$ $(V_o = -0.76$ V)

a. What is the voltage from the Ag–Zn battery resulting from the following electrochemical reaction under standard conditions?

$Zn + 2Ag^+ \rightarrow Zn^{++} + 2Ag$

b. The battery is being drained at a constant current of 10 amps. The zinc electrode of this battery weighs 65 g. Given that zinc dissolution is the factor controlling battery life, show that the battery life is about 5.5 h. The atomic weight of Zn is 65, and $N_A = 6 \times 10^{23}$, $e = 1.6 \times 10^{-19}$ C.

3. It is possible to make a battery based on the oxidation of aluminum. The relevant half-cell reactions follow, along with the standard electrode potentials:

$Al \rightarrow Al^{+++} + 3e$ $(V_o = -1.66$ V)

$4(OH)^- \rightarrow O_2 + 2H_2O + 4e^-$ $(V_o = 0.40$ V)

a. What is the voltage from this battery resulting from the following electrochemical reaction under standard conditions?

$4Al + 3O_2 + 6H_2O \rightarrow 4Al(OH)_3$

b. Aluminum is being consumed in this battery. Given that the atomic weight of Al is 27, $N_A = 6 \times 10^{23}$, and $e = 1.6 \times 10^{-19}$ C, calculate how long 1 g of aluminum will last when the current is 1 amp.

4. Corrosion in metals and alloys:

a. Grain boundaries are more susceptible to corrosion than the grain matrix. Explain.

b. Explain why pure metals in general are more corrosion resistant than impure ones.

5. Drink for thought:

Aluminum used in beverage containers is subjected to corrosion when exposed to the liquid. The industry attempts to solve this problem by coating the inside of the can with a polymer barrier coating. However, pinholes and cracks are always present in the coating due to processing. The industry standard is that, under some standard testing conditions, if the total corrosion current is less than 0.04 mA, it is considered to be acceptable.

Assume that (1) the same maximum corrosion current exists in a standard soft drink can, and (2) each can is on the shelf for 1 month before being consumed. If you drink one can of soft drink every day, show that you will ingest about 4 g of aluminum every year. The following information may be useful: The valence of aluminum is 3, and its atomic weight is 27 amu (1 amu = 1.67×10^{-27} kg).

6. You are given the following electrode reactions:

$Cu \rightarrow Cu^{2+} + 2e$ $(V_o = +0.34$ V)

$Al \rightarrow Al^{3+} + 3e$ $(V_o = -1.66$ V)

a. What is the standard voltage from a Cu–Al battery resulting from the following electrochemical reaction?

$2Al + 3Cu^{2+} \rightarrow 2Al^{3+} + 3Cu$

b. The battery is being drained at a constant current of 10 amps. Calculate the rate of consumption of aluminum in atoms per second.

7. The free energy released in the following reaction under standard conditions is 1670 kJ/mol of Al_2O_3:

$2\,Al + 3/2\,O_2 \rightarrow Al_2O_3$

a. Assuming that a fuel cell is made using this reaction, calculate the voltage output from this fuel cell under standard temperature and pressure conditions.

b. Assuming that the fuel cell is 5% aluminum by weight and that the energy conversion efficiency is 80%, calculate the energy storage capacity of this fuel cell in the form of watthours per kilogram.

8. A metal under stress tends to corrode more easily than one without stress. This means that the electrode potential for this metal is more negative by an amount ΔV. Estimate ΔV in two steps as follows:
   a. For a metal with an elastic modulus of E and elastic strain ε, show that the elastic strain energy per mole is equal to $E\varepsilon^2 A/2\rho$, where A is the atomic weight and ρ the mass density. (Hint: elastic strain energy per unit volume is equal to $E\varepsilon^2/2$.)
   b. The electrode reaction for this metal M is $M \to M^{n+} + ne$; express ΔV in terms of E, ε, A, ρ, n, and e, where e is the electron charge. (Hint: Recall the relationship between energy per mole and electrode potential.)
9. Consider two half-cells connected together so that one acts as cathode, the other as anode (corroding side). The following data are given, where subscripts c and a stand for cathode and anode, respectively:
   $V_{o,c}$ (standard potential for the cathode) = 0.2 V
   $\beta_c = -0.1$
   $J_{o,c} = 1$ nA/cm^2
   $V_{o,a}$ (standard potential for the anode) = -0.5V
   $\beta_a = +0.1$
   $J_{o,a} = 10$ nA/cm^2
   a. At steady state, the two half-cell electrode potentials are equal (i.e., $V_c = V_a$), where $V_c = V_{o,c} + \beta_c \log_{10}(J/J_{o,c})$ and $V_a = V_{o,a} + \beta_a \log_{10}(J/J_{o,a})$. Show that the steady-state corrosion current density J is 1×10^{-5} A/cm^2.
   b. From the preceding result, show that the corrosion potential (V_c or V_a) at steady state is equal to -0.2 V.
   In this problem, we assume that the corrosion current is limited by activation polarization.
10. Take an aqueous solution of sodium chloride with two immersed iron electrodes biased using a 6-V battery. Gas bubbles form on the negative electrode, while a reddish substance precipitates from the positive electrode. Discuss the reactions occurring at these electrodes.
11. Consider an iron wire of length 1 cm and diameter 1 mm, with electrical resistivity of 10^{-5} ohm-cm. One end of the wire is plastically deformed such that the strain energy density is 10 J/cm^3.
    a. For the Fe/Fe^{2+} half-cell reaction, calculate the change in standard electron potential in volts due to the strain energy. The atomic number of iron is 56, and its density is 8 g/cm^3.
    b. Calculate the corrosion current, assuming that it flows from one end of the wire to the other.
12. Alumina does not dissolve in saltwater, but dissolves in a concentrated hydrochloric acid solution. In the latter case, the dissolution rate is much slower than that for aluminum. Explain these observations.

9 Magnetic Properties

9.1 FLASHLIGHT WITHOUT BATTERIES

How often have you picked up a flashlight and found that the battery was dead? Of course, this happened at a time when you needed it most, and no spare batteries were around. One solution is to have a flashlight that converts mechanical to electrical energy and stores the energy in a capacitor,* as shown in Figure 9.1.

In this device, the back-and-forth motion of the magnet induces the flow of an alternating current in the solenoid coil. Appropriate electronics are used to convert this alternating current into a direct current, which then charges the capacitor (or a rechargeable battery). Let us estimate how much energy we can obtain by passing the magnet through the solenoid coil in 0.2 s from one end to the other (or 0.1 s to the midpoint of solenoid). The induced voltage V in volts is given by:

$$V = -\frac{d\phi}{dt}$$

$$= -\frac{d}{dt}(NBA),$$

(9.1)

where

ϕ = the total magnetic flux passing through the solenoid
N = the number of turns in the solenoid coil
B = the magnetic induction in teslas
A = cross section of the solenoid coil in square meters
t = the time

Let us assume the following parameters: maximum magnetic induction $B = 1$ T (1×10^4 G), radius of the solenoid coil = 10^{-2} m (1 cm), and number of turns in the solenoid coil = 1000. For the purpose of this calculation, assume further that when the magnet is at the end of the travel, the magnetic field experienced by the solenoid goes to zero. Under these conditions, the induced voltage is equal to:

$$\frac{\Delta\phi}{\Delta t} = \frac{1000 \times 1 \times \pi \times (0.01)^2}{0.1} = 3.14 \text{ V}.$$

Assume that the electrical resistance R of the coil is 50 ohms. The energy stored for the entire excursion of the magnet from one end to the other is given by:

$$\frac{V^2}{R}2\Delta t = \frac{(3.14)^2}{50}0.2 \approx 0.04 \text{ J}.$$

---

* I first found a commercial product like this in Hong Kong in 2002. Unlike rechargeable batteries, a capacitor can be charged and discharged tens of thousands of times without memory loss or other adverse effects.

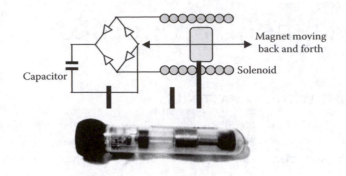

FIGURE 9.1 Conversion of mechanical energy to electrical energy.

Repeating this process 100 times produces 4 J, which is sufficient to power a small light-emitting diode (10 mW) for a few minutes. In this calculation, we ignore the fact that the voltage is not constant throughout.

This is the same principle used in electricity generation in power plants and in alternators and magnetos* used in cars and planes: Interception of magnetic flux lines by a moving electrical conductor results in a voltage and hence passage of an electrical current across the conductor. The magnitude of this voltage V is proportional to the magnetic field B and speed. Since power increases as V^2, the mechanical-to-electrical energy conversion efficiency increases as B^2. Therefore, there is significant interest in developing strong magnets for efficient electrical power generation.**

9.2 TINY MAGNETS FOR DATA STORAGE

Another major application of magnetic materials is in magnetic data storage. In this case, we work with small magnets. How small? A standard hard-disk drive in 2005 can store about 80 Gbits/in.². Therefore, each bit occupies an equivalent area of ~90 nm × 90 nm (the actual bit is rectangular with a typical aspect or width-to-length ratio of between 4 and 8). The goal of the disk-drive industry is eventually to achieve an areal storage density of 1 Tbit/in.², which translates into roughly 25 nm × 25 nm per bit — really small magnets indeed. These magnets have to be quite strong for two reasons: improved read-back signal and better thermal stability. Whether a given data bit reads one or zero depends on the orientation of the magnetic domain (as it is properly called) — that is, which way the north pole of the tiny magnet points. The energy ΔE required to flip the orientation of the magnetic domain is given by:

$$\Delta E = KV, \tag{9.2}$$

where V is the domain volume and K, the anisotropy constant, is material dependent. The frequency f at which the domain flips or reverses from one orientation to another due to thermal excitation is given by:

---

* The engine of a typical general aviation aircraft is energized by magnetos. In operation, each magneto is an open-circuit solenoid coil spun by the engine between magnets and generates a high voltage using the same principle as the example shown here. The generation of this high voltage (which is applied to the spark plugs) is timed precisely to ignite the fuel–air mixture at the completion of the compression cycle for a typical four-cycle internal combustion engine. For redundancy, each cylinder of the engine is energized by two magnetos. As long as the engine is running, no external electrical power is needed, and the aircraft will fly even when there is complete electrical failure.

** The inverse process, the conversion of electrical energy to mechanical motion, can be obtained using electric motors. The use of strong magnets is also important for efficient conversion. However, as discussed at the end of the chapter, an electrical conductor moving in a magnetic field results in a viscous drag force, which increases as the square of the magnetic field, thus limiting the maximum speed of the motor. Special design is incorporated in such motors to allow variable speed operations.

$$f = f_o \exp\left(-\frac{KV}{k_B T}\right),$$ (9.3)

where k_B is the Boltzmann constant, T absolute temperature, and f_o the attempt frequency, which is usually assigned a value of 1×10^9/s.

   Clearly, this reversal frequency increases with decreasing domain volume. This means that when it is sufficiently small, a data bit may spontaneously change from zero to one or vice versa; this is clearly unacceptable for a data storage device. This poses a physical limit, known as the superparamagnetic limit, on the maximum magnetic storage density.

EXAMPLE

Consider a magnetic domain with $K = 5 \times 10^5$ J/m^3 and a volume V of $10 \times 10 \times 10$ nm^3. Estimate the reversal frequency at room temperature (300 K). How will the reversal frequency change when the volume is reduced by a factor of two?

Solution

The domain volume is equal to 1×10^{-24} m^3. Using Equation (9.3), we can determine the reversal frequency as follows:

$$f = 10^9 \exp\left(-\frac{5 \times 10^5 \times 10^{-24}}{1.38 \times 10^{-23} \times 300}\right) = 10^9 \exp(-121) \approx 10^{-43}/s.$$

This reversal frequency is for one domain. For a disk with 10^{12} domains, the reversal frequency is just $10^{12} \times 10^{-43} = 10^{-31}$/s. This means that one reversal occurs every 10^{31} s (cf. the estimated life of the universe of 5×10^{17} s) somewhere on the disk. When the volume is reduced by a factor of two, the exponential term becomes $\exp(-60.4)$, and the reversal frequency f is given by:

$$f = 10^9 \exp(-60.4) \approx 10^{-17}/s.$$

Therefore, halving the domain volume increases the reversal frequency by 26 orders of magnitude. In the latter case, if the disk has 10^{12} such domains, one reversal will occur every 10^5 s (~28 h) somewhere on the disk. This is a significant error rate.

   From 1954 to 2006, the magnetic storage density increased exponentially from about 2 kbits/in.^{2} to almost 100 Gbits/in.^{2}.* The preceding discussion indicates that this exponential growth cannot continue forever because there is a physical limit to the magnetic storage density, due to the thermally excited reversal of magnetic domains. It is not clear how the technology and the industry as a whole may evolve if and when we reach this limit.

---

* Magnetism is a quantum mechanical phenomenon appearing in the macroscopic world. Since magnetic domains in magnetic storage are in the nanometer range, it is fair to say that magnetic data storage is one of the best examples of how quantum mechanics and nanotechnology work together to affect our everyday lives. In 1990 when it started to be common-place to include a hard drive in a personal computer, the cost was about $1 per megabyte. In 2005, the same 1-Mbyte hard drive storage costs only 0.1 cent and is dropping fast. The magnetic media layer that stores the information is about 15–20 nm thick, covered by a 2–3 nm protective overcoat and a monomolecular lubricant layer.

9.3 MAGNETISM FUNDAMENTALS AND DEFINITIONS

In this chapter, we will explore several practical magnetism topics related to power generation, data storage, and diagnostics. Before we do so, we will go over some fundamentals and a few definitions. In this chapter, we use practical meter-kilogram-second (m.k.s.) units. Where appropriate, the same quantities will be expressed in centimeter-gram-second (c.g.s.) units.

9.3.1 MAGNETIC FIELD

A magnetic field can be generated by an electrical current. An infinite solenoid with n turns per meter produces a uniform magnetic field H inside the solenoid given by:

$$H = nI, \tag{9.4}$$

where I is the current (ampere) passing through the solenoid. The unit of magnetic field H is therefore ampere-turns per meter. In c.g.s. units, H is in oersteds (1 ampere-turn/meter is equal to $4\pi \times 10^{-3}$ Oe; the Earth's magnetic field in midlatitudes is around 0.3–0.5 Oe*).

EXAMPLE

How can a solenoid be used to generate a magnetic field of 10^4 Oe?

Solution

$H = 10^4$ Oe $= 10^4/(4\pi \times 10^{-3}) \sim 8 \times 10^5$ amp-turns/meter. Any combination of nI equal to 8×10^5 will work. For example, if $n = 1 \times 10^4$ turns/meter, then current $I = 80$ amps.

Note that for an infinite solenoid, the magnetic field does not depend on the radius of the solenoid. In practice, Equation (9.4) holds near the center of a solenoid when the length/radius is ≥ 10.

9.3.2 MAGNETIC MOMENT AND MAGNETIZATION

Magnetism arises from electrons having nonzero net angular momenta due to their combined orbital motions and intrinsic spins. In the early days, a magnet was assumed to be analogous to an electrostatic dipole (i.e., there were separate magnetic charges or monopoles). The strength of a magnet (or magnetic dipole) was given by the magnetic moment, defined as the magnetic charge times the distance between the magnetic charges. Based on this definition, the magnetic field at some distance from the magnet could be calculated.

Since separate magnetic charges have yet to be found, it is intellectually more satisfying to express the magnetic moment in some other way. As noted earlier, an electrical current produces a magnetic field; a current I flowing in a loop of area A behaves like a magnet with a magnetic moment m equal to IA (Figure 9.2). Therefore, the unit of magnetic moment is ampere-square meters. One electron spin produces an equivalent magnetic moment of 9.27×10^{-24} amp-m^2. This is known as one Bohr magneton (μ_B).** A real magnet consists of many such Bohr magnetons. In this case, the magnetic strength of a material is measured by the magnetic moment per unit volume and is known as magnetization M, which has the same unit as H.

---

* It is generally believed that the motion of the Earth's molten core, rich in iron, is the source of this magnetic field.
** For an electron moving in the 1s orbital of a hydrogen atom, the radius of the orbit is 0.053 nm, and the electron speed is (1/137) times the speed of light. Using this information, it can be shown that the electron current times the area enclosed by the 1s orbit is equal to one Bohr magneton.

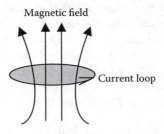

FIGURE 9.2 Magnetic field produced by a current loop.

9.3.3 MAGNETIC INDUCTION OR FLUX DENSITY

A piece of unmagnetized iron, for example, consists of electron spins organized into domains. These domains are randomly distributed in six equivalent <100> directions so that the net magnetization for the sample is zero.* Within each domain, the spins point in the same direction. Now, let us put this piece of unmagnetized iron inside an energized solenoid. The energy of magnetic dipoles E_{ms} in a magnetic field H in free space is given by:

$$E_{ms} = -\mu_o \, m \, H \cos \theta, \tag{9.5}$$

where

μ_o = a constant for unit conversion ($4\pi \times 10^{-7}$), known as permittivity of free space
m = the magnetic moment
H = the magnetic field
θ = the angle between the orientation of the magnetic moment and the magnetic field

To lower the magnetostatic energy E_{ms} of the system, magnetic moments prefer to line up along the direction of the applied magnetic field — that is, $\theta = 0$. As a result, magnetic domains in the favorable direction grow, and the net magnetization increases (Figure 9.3). This in turn increases the magnetic field inside the iron, which makes it easier to line up the spins. Instead of using the term magnetic field to describe the field inside the material, we use the term *magnetic induction* B. We can visualize B as a measure of the density of magnetic flux lines in the material. The relationship among B, H, and M is given by:

$$B = \mu_o (H + M) . \tag{9.6a}$$

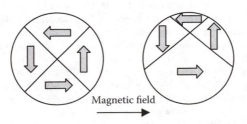

FIGURE 9.3 Magnetic domain growth in the presence of an applied field.

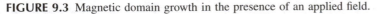

* These are known as easy directions. For nickel, <111> are the easy directions. For cobalt, <0001> are the easy directions. The preference of the magnetization vector to be along specific crystallographic directions is known as crystalline anisotropy. Two other factors induce anisotropy in magnetism: stress and shape.

Therefore, contributions to the magnetic induction or flux density come from two terms: the applied field ($\mu_o H$) and the magnetization ($\mu_o M$). Occasionally, we use the term *magnetic field* for B in free space, air, or nonmagnetic materials. The unit for B is teslas or webers per square meter. In c.g.s. units, B is in gauss (1 T = 10^4 G).* In general, the energy of a magnetic moment in a magnetic induction or flux density B is given by:

$$E = -mB\cos\theta,\qquad(9.6b)$$

where E is the energy, m the magnetic moment, and θ the angle between the magnetic moment and magnetic induction. As a result, an alternate unit for magnetic moment is joules per tesla. Note that Equation (9.5) is a special case of Equation (9.6b); in free space, $B = \mu_o H$. When M is proportional to H so that $M = \chi H$ (χ is known as the susceptibility),

$$B = \mu_o(H + \chi H)$$
$$= \mu_o(1+\chi)H\qquad(9.6c)$$
$$= \mu\mu_o H,$$

where μ (=1 + χ) is known as the permeability of the material.

9.3.4 SATURATION MAGNETIZATION AND FORCE OF ATTRACTION

Experimentally, the maximum or *saturation magnetization* M_{sat} for Ni is equal to 5.1×10^5 amp/m. From this information, we can calculate the average magnetic moment per Ni atom m as follows:

$$m = \frac{M_{sat}}{\text{number of } Ni \text{ atoms/volume}}$$
$$= \frac{5.1\times10^5}{9.1\times10^{28}}$$
$$= 5.6\times10^{-24} \text{ amp}-\text{m}^2$$
$$= \frac{5.6\times10^{-24}}{9.27\times10^{-24}}\mu_B$$
$$= 0.6\mu_B.$$

---

* Amid our discussion of units and conversion, I cannot help but cite a well-known aviation incident that occurred on July 23, 1983. This involved Air Canada Flight 143. The Boeing 767 aircraft departed Montreal and made a scheduled stop at Ottawa before continuing to Edmonton. Because of the failure of the fuel processor, the flight crew relied on ground personnel at Ottawa to obtain a manual measurement of the fuel onboard. The crew was told that 11,430 L of fuel was onboard. Using the conversion factor of 1.77, the flight crew thought that they had 20,200 kg of kerosene, the fuel quantity needed to complete the 1700-mile trip to Edmonton plus reserve. Instead, they only had 9144 kg of fuel (the correct conversion is 1.77 lb/L or 0.8 kg/L). The plane ran out of fuel about 2 hours into the flight. The engine-driven generators stopped delivering power — lights went out, and all advanced avionics (including the critical flight-management computer) stopped working, leaving only a 24-V NiCd battery to operate the radio and standby instruments. More important, the electrically powered hydraulic system used to drive all primary control surfaces (ailerons, elevators, and rudders), flaps, speed brakes, and landing gears also stopped functioning. Captain Robert Pearson and First Officer Maurice Quintal were flying the heaviest glider with fewer instruments than Charles Lindbergh. This would have meant certain disaster. Fortunately, the captain was a proficient glider pilot, and the plane was within gliding distance of an abandoned RCAF airfield at Gimli. More important, the Boeing 767 was equipped with a ram air-powered turbine that provided just enough electrical power to run the primary control surfaces. In the end, the crew made a successful emergency landing with no deaths or serious injuries. The plane was eventually repaired. Known as the Gimli Glider, it was still in service as of 2000.

Therefore, the average magnetic moment per Ni atom is 0.6 μ_B. The resulting magnetic induction or flux density B inside Ni due to M_{sat} is given by:

$$B = \mu_o M_{sat} = 4\pi \times 10^{-7} \times 5.1 \times 10^5 = 0.64 \text{ T.}$$

How can we cause all the tiny magnets in Ni to line up to give the saturation magnetization? We can do so by bringing a bar magnet next to Ni. As indicated by Equation (9.6b), the alignment of the magnetic moment in the same direction as the magnetic field or induction is a state of low energy (think of Ni as a sponge for the magnetic flux lines). As a result, the system (bar magnet + Ni) prefers to remain in this state — that is, Ni sticks onto the magnet, a common observation. Assume that the magnet has a cross-section area A and uniform magnetization M. The force of attraction F between the magnet and the soft magnetic material is given approximately by:

$$F = \frac{1}{2}\mu_o M^2 A. \tag{9.7}$$

This formula is strictly correct for an infinitely long magnet in contact with a material with infinite permeability. However, it works well with magnets in which the length is at least five times the width and with magnetic materials having permeability greater than 100.

EXAMPLE

Using a magnet with $M = 1 \times 10^6$ amp/m and a cross-section area of 100 cm², calculate the attractive force on a magnetic material with infinite permeability.

Solution

From Equation (9.7), we can write:

$$\frac{F}{A} = \frac{1}{2}\mu_o M^2 = 0.5 \times 4\pi \times 10^{-7} \times (10^6)^2$$

$$= 6.28 \times 10^5 \text{ N/m}^2.$$

At 100-cm² of cross-section area (i.e., $A = 10^{-2}$ m²), the attractive force F is given by:

$$F = 6.28 \times 10^5 \times 10^{-2}$$

$$= 6.28 \times 10^3 \text{ N.}$$

This force allows the magnet to hold up a weight of about 1400 lb and is the basis of using magnets to separate (magnetic) ferrous materials in recycling operations.

9.4 DIAMAGNETIC AND PARAMAGNETIC MATERIALS

We will take a short digression to talk about two types of weak magnetic responses. The first type is diamagnetism. Consider a material (e.g., sodium chloride) in which all atoms have paired electrons. We know from common experience that sodium chloride is nonmagnetic. In an external magnetic field, such a material responds by producing a weak magnetization M oriented in the *opposite* direction. This is known as a diamagnetic response. In this case, the ratio M/H is defined as diamagnetic susceptibility and is typically in the 10^{-5} range for most materials.

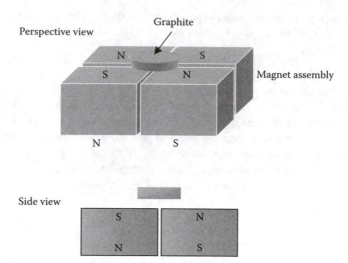

FIGURE 9.4 Magnetic levitation of a graphite flake.

To demonstrate this diamagnetic response, obtain four strong pancake magnets and assemble them in a square pattern as shown in Figure 9.4. Carefully place a small piece of graphite, a diamagnetic material, near the center of the magnet assembly.* Note the levitation of the sample above the magnets (Figure 9.4). How does the levitation occur? For a diamagnetic material, the magnetization M is given by:

$$M = \chi\, H. \tag{9.8a}$$

Here, χ is the diamagnetic susceptibility, which is a negative number. This means that the magnetization vector is opposite to the direction of the magnetic field H. The energy of a diamagnetic material in magnetic field H is:

$$E = -\frac{1}{2} MH = -\frac{1}{2}\chi H^2. \tag{9.8b}$$

The levitation force F acting on the sample is given by:

$$F = -\frac{dE}{dz} = -\chi H \frac{dH}{dz}, \tag{9.8c}$$

where z is the vertical direction as shown. When this quantity is equal to the weight of the graphite sample at a given height above the magnet surface, levitation occurs as observed. The reason that this works well for graphite is due to its relatively high diamagnetic susceptibility (see Table 9.1).**

On the other hand, there are materials with unpaired electrons and hence permanent magnetic dipole moments. In the absence of an external magnetic field, these magnetic dipoles are randomly

---

* Pencil lead should work, but pyrolytic graphite works best. The square array of magnets is to provide for stable levitation. As demonstrated in this experiment, levitation can be performed without superconductors.
** A superconductor is a perfect diamagnet; in the presence of an external field H, it creates a supercurrent to counteract so that the magnetic induction B is zero. This implies that $H + M = 0$. The diamagnetic susceptibility is then equal to M/H = −1. Therefore, the magnetic levitation force acting on a superconductor is about 100,000 times greater than that on most diamagnetic materials, and about 6,000 times greater than that on graphite.

TABLE 9.1
Diamagnetic Susceptibility
for Several Selected Materials

Materials	$-\chi$ (in units of 10^{-6})
Graphite	160
Pyrolytic graphite ($\perp$)	450
Pyrolytic graphite ($\parallel$)	85
Bismuth	170
Gold	34
Water	9

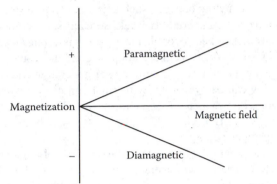

FIGURE 9.5 Schematic diagram showing the variation of saturation magnetization of a ferromagnet versus temperature.

oriented. When a magnetic field is applied, these dipoles line up, thus producing a net magnetization in the *same* direction as the external field. This phenomenon is known as paramagnetism. Paramagnetic susceptibility for most materials at room temperature ranges between 10^{-6} and 10^{-3}. Figure 9.5 contrasts the difference between diamagnetic and paramagnetic materials in how the magnetization changes with an applied magnetic field. Note the sign difference in magnetization.

Question for discussion: How does paramagnetic susceptibility *M/H* depend on temperature? (Hint: Think about the alignment of magnetic dipoles along the magnetic field when they are excited thermally.)

Without an external magnetic field, diamagnetic and paramagnetic materials have zero magnetization. Because of their low susceptibilities, the magnetization of these materials under moderate fields is quite small. As a result, they are considered to be nonmagnetic.

Nonmagnetic materials are used in scientific instruments that require low internal magnetic fields to function properly — for example, low-energy electron spectrometers, compass mounting fixtures, certain components used in magnetometers, etc. Paramagnetic materials are used routinely to obtain ultralow temperatures using a method known as adiabatic demagnetization. The first step of this method is to align magnetic moments of the paramagnetic material held at some initial (low) temperature T_{init} using a strong magnetic induction B. The ordered arrangement of magnetic moments corresponds to a state of low entropy. Statistical thermodynamics shows that the entropy of the system depends only on B/T, where T is the absolute temperature. Now reduce the magnetic induction to some small residual value under adiabatic conditions (i.e., no heat exchange with the outside world, $dQ = 0$). Since entropy change is equal to dQ/T, $dQ = 0$ means that the entropy is constant. If entropy is constant, then B/T is constant. Therefore, we have

$$\frac{B_{final}}{T_{final}} = \frac{B_{init}}{T_{init}}. \tag{9.9}$$

For $B_{init} = 1$ T and $B_{final} = 10^{-5}$ T, the temperature can be dropped from 4 K to 40 μK using this method.

9.5 MAGNETIC MATERIALS: FERROMAGNETISM AND ANTIFERROMAGNETISM

Let us return to the discussion of strongly magnetic materials. At this point, it is fair to raise the following question. A paramagnetic substance consists of atomic scale magnets, as does a permanent magnet. What prevents a paramagnetic substance from becoming a magnet? The difference lies in the strength of interaction or coupling between such adjacent electron spin moments. The coupling of spin moments in paramagnetic materials is weak, so each spin moment behaves more or less independently. In materials such as Fe, Co, or Ni, the coupling is strong and favors parallel alignment of adjacent spins. This is known as ferromagnetic coupling, and the resulting materials are known as ferromagnets. Ferromagnetic materials are used in power generation and data storage applications.

Ferromagnetic coupling can be overcome by thermal agitation; a magnet loses its magnetism above a critical temperature known as the Curie temperature (Figure 9.6). The Curie temperatures for Fe, Co, and Ni are 768, 1120, and 335°C, respectively. Above the Curie temperature, a ferromagnet becomes a paramagnet.

In ferromagnets, the coupling of spin moments favors parallel alignment of adjacent spins. In another class of materials known as antiferromagnets, the coupling favors antiparallel alignment of adjacent spins. For antiferromagnets (e.g. NiO), the magnitude of adjacent spins is the same so that the net magnetic moment is zero (Figure 9.7). In other cases such as cubic ferrites MFe_2O_4 (where M is one of several metallic elements),* the total up-moments are not equal to the total down-moments, so the net magnetic moment is nonzero. This is known as ferrimagnetism and is schematically represented in Figure 9.7.

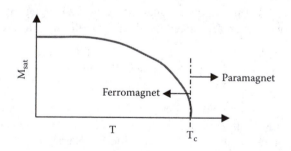

FIGURE 9.6 Saturation magnetization of a ferromagnetic material versus temperature.

---

* The best known ferrite is magnetite, Fe_3O_4. In one unit cell of magnetite, there are eight Fe^{2+} and 16 Fe^{3+} ions. All eight Fe^{2+} ions sit in octahedral sites and have parallel spins, each contributing four Bohr magnetons. Half of the Fe^{3+} ions sit in octahedral sites and half in tetrahedral sites with antiparallel spins. As a result, the Fe^{3+} ions do not contribute to the overall magnetic moment of magnetite. Given the unit cell size of 0.84 nm, it can be shown that the saturation magnetization of magnetite is equal to 5×10^5 A/m — almost identical to that of nickel.

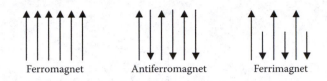

FIGURE 9.7 Comparing ferro-, antiferro-, and ferrimagnetism.

9.6 MAGNETIC MATERIALS FOR POWER GENERATION

We use ferromagnetic materials in two ways for power generation. One is the generation of induced voltage through the Faraday effect as described earlier in this chapter. The other is in transformers that convert AC (alternating current) voltages of one magnitude to another. Each application requires ferromagnetic materials with different properties, as measured by the $B–H$ or $M–H$ hysteresis curves described next.

Let us start with an initially nonmagnetic ferromagnet, in which the distribution of magnetic domains is such that the net magnetization is zero. We then measure the variation of magnetization M as a function of H as follows (Figure 9.8a):

- Along path 1–2, as the magnetic field increases, so does the magnetization. At sufficiently large fields, the magnetization reaches a saturation value M_{sat}. The ease with which saturation magnetization is attained depends on the mobility of magnetic domain walls. Dislocations, grain boundaries, and boundaries between phases or precipitates interfere with the motion of domain walls.
- Along path 2–3, the magnetic field decreases. The magnetization does not decrease along the original path 1–2. This is known as hysteresis. When the external field is zero, a nonzero residual magnetization known as remanence M_r remains (made possible by the strong coupling between neighboring spins as discussed in the preceding paragraph).

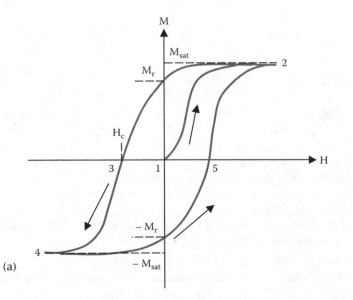

(a)

FIGURE 9.8 (a) A typical $M–H$ hysteresis curve for a ferromagnet; (b) demagnetization by shrinking the hysteresis loop; (c) $M–H$ curves for hard and soft magnets.

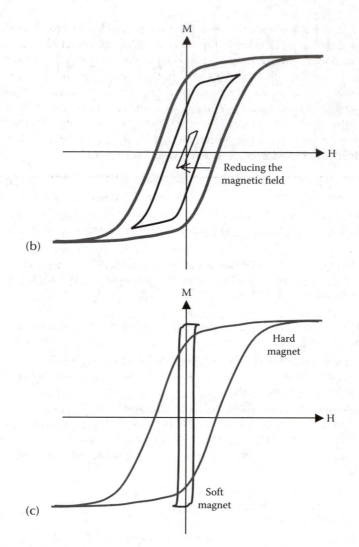

FIGURE 9.8 (continued)

- It takes a magnetic field of strength H_c in the opposite direction to reduce the magneti-zation to zero. H_c is known as the coercivity, which depends on the mobility of domain walls and hence the microstructure of the material.
- Along path 3–4, with a further increase in the magnetic field in the opposite direction, the magnetization reverses and eventually saturates.
- The changes along paths 4–5 and 5–2 are analogous to paths 2–3 and 3–4, except that the direction is reversed.
- Finally, note that the area enclosed by the M versus H hysteresis curve is equal to the energy dissipated when one loop is executed.

How to Demagnetize Objects

A demagnetizing or degaussing coil is often used to demagnetize a given object, whether it is a TV set, a screwdriver, or a component used in special scientific instruments. A typical degaussing coil is a solenoid, about 30 cm in diameter and 2 cm thick, connected to an AC power source. The object to be demagnetized is first placed at the center of the degaussing coil. Then the degaussing coil is slowly moved away from the object (2 m away or greater), at which point the power to the degaussing coil is turned off. The object is now demagnetized.

During each AC cycle (which is 1/60 s long in the United States), the object undergoes a full hysteresis cycle induced by the degaussing coil. By moving the degaussing coil away from the object, the driving magnetic field is reduced, as is the remanent or residual magnetization. In other words, the size of the hysteresis loop shrinks (Figure 9.8b). Eventually, when the degaussing coil is far away from the object, the driving magnetic field is close to zero, and the remanent magnetization approaches zero as well.

For electrical power generation, we need to use strong permanent magnets — that is, large M_r and large H_c at the operating temperature. Larger M_r means higher flux density to be intercepted by the moving conductor, giving a larger induced voltage and hence higher mechanical-to-electrical energy conversion efficiency. Larger H_c means that the magnet stays as a permanent magnet without being accidentally demagnetized. Materials with these properties are known as hard magnets (Figure 9.8c). By virtue of their large M_r and H_c values, it takes a lot of energy to demagnetize hard magnets. Examples of hard magnetic materials include Alnico 8 ($M_r = 6.0 \times 10^5$ A/m and $H_c = 1.2 \times 10^5$ A/m), Co_5Sm ($M_r = 7.3 \times 10^5$ A/m and $H_c = 7.2 \times 10^5$ A/m), and $Nd_2Fe_{14}B$ ($M_r = 9.2 \times 10^5$ A/m and $H_c = 8.5 \times 10^5$ A/m).

As indicated by Equation (9.6a), a strong magnet means large saturation or remanent magnetization, which depends on the intrinsic electronic structure of the material. Ferromagnetic materials made of multiple components have rather unpredictable properties, including the Curie temperature and saturation magnetization. For example, at room temperature, M_{sat} (Fe) = 1.71×10^6 amp/m and M_{sat} (Co) = 1.42×10^6 amp/m. Addition of 30 w/o of cobalt to iron increases the Curie temperature from ~770 to 950°C, which is not surprising (since the Curie temperature of cobalt is higher than iron). However, what is surprising is that this alloy has a higher M_{sat} (~1.87×10^6 amp/m) than the pure components.

In addition to the composition of the magnet, coercivity depends on the microstructure. Imperfections such as dislocations, grain boundaries, and precipitates impede domain wall motion and thus increase coercivity. Another factor controlling coercivity is the shape of the magnetic material. Whenever something is magnetized, an internal demagnetizing field aids domain reversal. One way to think about this demagnetizing field is as follows. Inside a bar magnet, the magnetic field is in the direction of the north pole to the south pole of the magnet. This field provides the driving force to line up magnetic domains so that the north pole of the domains is pushed towards the south pole of the bar magnet, thus acting to demagnetize the magnet. The strength of this demagnetizing field depends on the shape of the material. It is stronger for a thin-disk magnet and weaker for a long-bar magnet.

In transformer applications (Figure 9.9), an AC voltage is applied to the primary solenoid coil wrapped around a ferromagnetic core. The resulting current produces a magnetic field and hence magnetization inside the core. The time-varying magnetic flux induces a voltage on the secondary coil. For an ideal transformer, the magnitude of the AC voltage appearing on the secondary coil is

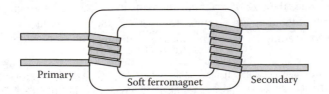

FIGURE 9.9 Schematic diagram of a transformer.

proportional to the ratio of the number of turns in the secondary coil to that of the primary. To minimize energy loss, the area enclosed by the hysteresis loop must be as small as possible.

Generally, it is desirable for the transformer core to have a large saturation magnetization M_{sat}, obtainable at a low magnetic field and small coercivity H_c. This usually means that the hysteresis loops are narrow and rectangular. Materials with these properties are known as soft magnets (Figure 9.8c). Common soft magnetic materials include Fe with 3 w/o Si ($M_{sat} = 1.6 \times 10^6$ A/m, $H_c = 40$ A/m, and energy loss per cycle = 40 J/m^3) and iron with 45 w/o Ni (known as 45 Permalloy with $M_{sat} = 1.3 \times 10^6$ A/m, $H_c = 10$ A/m, and energy loss per cycle = 120 J/m^3).

More recently, amorphous magnetic alloys of different compositions have been used because of their magnetic softness and excellent corrosion resistance. The same principle of inductive coupling is used to recharge batteries implanted in human bodies (e.g., in pacemaker applications). AC power is coupled through the skin via an internal solenoid wound around a soft magnetic core. No surgeries or invasive techniques are needed to recharge the batteries. In this case, it is important that these materials be properly encapsulated or have excellent corrosion resistance.

Question for discussion: Why are amorphous alloys magnetically soft? (Hint: Certain microstructures impart "magnetic hardness." What are these features? Are some of these features absent in amorphous alloys?)

9.7 MAGNETIC MATERIALS FOR DATA STORAGE

As discussed in Section 9.2, one of the major concerns for high-density magnetic storage is the thermal stability of the magnetic storage media — that is, a given magnetic domain changes the direction of magnetization spontaneously due to thermal excitation. For cobalt-based magnetic media, magnetization can only have two directions (along the $\pm c$ axis of the unit cell). In the absence of an external magnetic field, the activation energy required to flip a domain of volume V between these two directions is equal to KV, where K is known as anisotropy constant given by:

$$K = \frac{1}{2}\mu_o M_r H_c, \qquad (9.10a)$$

where μ_o is the permeability of free space, M_r the remanence, and H_c the coercivity. For a magnetic domain of area A and thickness t, the activation energy ΔE is then given by:

$$\Delta E = \left(\frac{\mu_o}{2} M_r t H_c\right) A . \qquad (9.10b)$$

To minimize the superparamagnetic effect discussed earlier, we must make ΔE much greater than the thermal energy $k_B T$. For a given magnetic domain or bit size A, we can accomplish this goal by maximizing $M_r t H_c$, which is therefore considered to be the figure of merit in designing magnetic media for maximum storage density. Note that we cannot make the coercivity so large as to create problems in writing data onto the hard disk.*

In our quest to reduce the size of magnetic domains in disk drives (to increase the areal storage density), we also reduce the magnetic signal. Before 1992, the same inductive coil used for writing bits was employed for reading as well. This method ran into sensitivity problems as we increased

---

* Data bits are written by an inductive coil embedded in the read–write head. During writing, a current pulse is sent to the inductive coil, generating a large magnetic field near the bottom of the read–write head in close proximity to the hard disk surface. The magnitude of this magnetic field is greater than the coercivity of the magnetic media and reorients the magnetic domain in the desired direction.

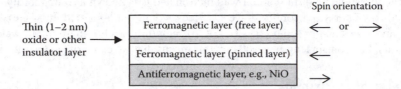

FIGURE 9.10 Schematic diagram of a typical GMR sensor used in disk drives.

the storage density to the gigabits per square inch regime. Today, we use a different sensing technique based on the phenomenon of magnetoresistance (MR) — the change of electrical resistance due to a magnetic field. Depending on how strong the magnetic field effect is on the electrical resistance, different names are used to describe this phenomenon: giant magnetoresistance (GMR) and colossal magnetoresistance (CMR). One popular GMR sensor, known as the spin valve, is shown schematically in Figure 9.10.

Because of its antiparallel coupling with adjacent spins, the antiferromagnetic layer forces the adjacent ferromagnetic layer to have a definite magnetization direction. Hence, this ferromagnetic layer is known as the pinned layer. The topmost ferromagnetic layer does not have this restriction, and its magnetization can point in any direction. Therefore, this layer is known as the free layer. Consider an external magnetic field (due to magnetic domains in the hard disk) being applied to this sensor so that it orients the spins in the free layer in the same direction as the pinned layer. Since electrons with the same spin tend to avoid each other, the electron–electron scattering probability is reduced. As a result, the electrical resistance decreases. Traditional materials such as Permalloys respond to magnetic fields in the 1000 amp/m range (~10 Oe) with resistance change ~ 1%. In comparison, the resistance change in a spin valve is ~10% under similar magnetic field conditions. It is interesting to note that the same concept as shown in Figure 9.10 is being explored as nonvolatile magnetic random-access memory (MRAM).

The final point to address in this section is speed. How fast can a bit be written on a hard disk? This may not be important for those who work with short text documents, but it means a great deal to individuals who constantly have to move massive amounts of data in and out of the hard drive — for example, in image and video processing. At low writing fields, magnetization occurs due to domain wall motion as shown in Figure 9.3.*

In electrically conducting magnetic materials (the magnetic media in hard disks and digital tapes are made of Co/Pt alloys), there are two limitations to domain wall velocity: eddy current damping and intrinsic damping. Eddy current damping is the same phenomenon discussed at the beginning of this chapter. As we apply a magnetic field with the intent to write a bit, the magnetic flux intercepted by the magnetic media layer increases. As a result, an induced electrical current flows within the magnetic layer. The direction of this induced current produces a magnetic field that opposes the applied magnetic field. It is this induced magnetic field that slows down the motion of the domain wall.

Eddy current damping is not important when the magnetic layer has high electrical resistance, as in thin films or high-resistivity ceramics. In this situation, the second factor takes over. Intrinsic damping is the retardation of electron spin rotation and is related to coupling with adjacent spin moments. To first order, the domain wall velocity v is related to the driving field H by the following equation:

$$v = C(H - H_o),\qquad(9.11)$$

---

* While it is beyond the scope of this text, it should be mentioned that, at high fields, domain rotation occurs, in which the magnetization is aligned with the direction of the applied field, even though this direction is not the easy magnetization direction for the material.

where H_o is the threshold field for domain wall motion and C is known as the mobility. Most media materials used for magnetic recording have C values ~ 1 m²/amp-s (~10^4 cm/s-Oe). With this C value and at an excess driving field $(H - H_o)$ of 10^2 amp/m (~1 Oe), the domain wall velocity is equal to 100 m/s. For domain length of 100 nm, this means that we can write at the rate of 10^9 bits/s.

9.8 MAGNETOSTRICTION

Magnetostriction refers to the change of dimension (expansion or contraction) when a ferromagnetic material is being magnetized. The magnetostriction strain for typical Fe/Co/Ni-based alloys is ~10^{-5} and can be as high as 10^{-3} for certain rare earth alloys. The fundamental mechanism is electron spin-orbit coupling in atoms; that is, the system wants to maintain the same relative orientation of an electron orbital and electron spin. Therefore, when an electron spin aligns with the external magnetic field, the electron cloud (orbital) distorts to follow the reorientation of the electron spin (Figure 9.11). For nonspherical electron clouds,* this distortion changes the interatomic distance and hence results in expansion or contraction of the material.

Analogous to piezoelectricity, magnetostriction also works in reverse. When a magnetic material is under stress (tension or compression), it attempts to reduce the imposed elastic strain by reorienting the magnetic domains. The net result is a change of magnetization. As can be seen from Figure 9.12, the effect can be substantial. This phenomenon has been exploited to generate and detect ultrasonic waves.

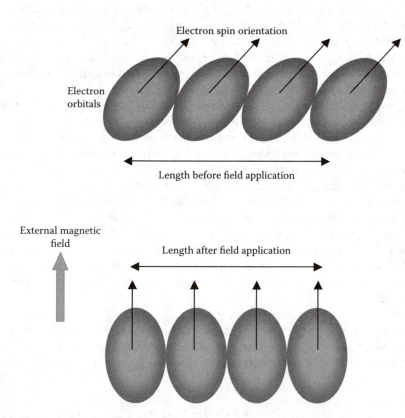

FIGURE 9.11 Illustration of the magnetostriction effect.

---

* Ferromagnetic materials are based on metals with d or f electrons. These electron orbitals tend to be highly directional.

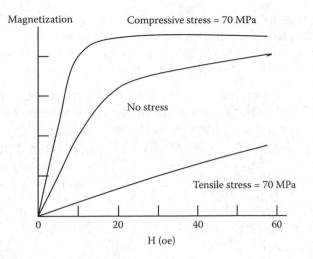

FIGURE 9.12 Effect of stress on the magnetization of nickel.

9.9 MEDICAL, SURVEYING, AND MATERIALS APPLICATIONS

Magnetic Resonance Imaging

Strong magnets (ranging from standard electromagnets to permanent and superconducting magnets) are used in a medical diagnostic technique known as magnetic resonance imaging (MRI).* This technique relies on the fact that the human body carries many protons (mostly in the form of water). Just like an electron, each proton has an intrinsic spin and behaves like a little magnet. Its magnetic moment is equal to $gm\mu_n$, where $g = 5.58$ for protons, $m = \pm\frac{1}{2}$, and μ_n is known as the nuclear magneton (5.05×10^{-27} amp-m^2), which is a factor of 1840 smaller than the Bohr magneton μ_B. In the presence of a magnetic field B, the proton spin can have two configurations: parallel or antiparallel to the field. The energy difference between these two configurations is $5.58\ \mu_n B$. At $B = 1$ T, $5.58\ \mu_n B \sim 2.82 \times 10^{-26}$ J. This energy is roughly equal to $6.6 \times 10^{-6}\ k_B T$ at 37°C, implying that an extra few parts per million of proton spins are aligned with the external field, and that is plenty.**

Consider what happens when we illuminate the aligned proton spins with an RF signal of frequency ν.*** Energy absorption occurs when one quantum of the electromagnetic wave energy is equal to $5.58\ \mu_n B$:

$$h\nu = 5.58\mu_n B \tag{9.12a}$$

* The first MRI examination performed on a human being took place on July 3, 1977, and required almost 5 h to produce one image. This machine is currently in the Smithsonian Institution.
** When stronger magnetic fields are used, the fraction of aligned proton spins — hence, the signal-to-noise ratio and spatial resolution in MRI — is increased. In 2001, the maximum field allowed in medical imaging of human patients in the United States was 2 T. Presumably, this field strength is considered to be safe. There have been numerous anecdotes from patients and workers operating MRI machines, ranging from feeling dizzy to having a disproportionate number of children of one sex versus the other.
*** In an actual MRI, such RF signals are directed to the body in pulses.

or

$$\nu = 4.2576 \times 10^7 B , \tag{9.12b}$$

where h is Planck's constant. Note from Equation (9.12b) that the RF frequency for energy absorption (sometimes known as the Larmor frequency) depends on the magnetic field or induction B. This is the basis of imaging in MRI systems.

After the input RF signal is switched off, the proton spins realign with the magnetic field with time constant T_1. If the input RF signal is oriented in such a way that the proton spins are momentarily aligned in a certain direction, the proton spin moments in that direction decay with another time constant T_2. We can describe the realignment process as the precession of proton spins about the magnetic field axis, resulting in the emission of RF signals of the same frequency. By providing a spatially varying magnetic field during the data acquisition process,* we can associate each point in space with an RF frequency, thus providing the capability to perform imaging. Since both time constants are related to how strongly the proton spins are coupled (or interact) with the environment, they provide the necessary mechanisms for image contrast between different types of tissues. For example, T_1(fat) = 240 ms, T_2(fat) = 60 ms, T_1(muscle) = 400 ms, and T_2(muscle) = 50 ms at $B = 0.2$ T.

9.9.1 HUNTING FOR OIL AND MINERAL DEPOSITS

A related but less sophisticated application of the same principle has been used in surveying for oil and mineral deposits. The basic idea is that large oil and mineral deposits create local magnetic anomalies. A sensitive instrument to measure the Earth's magnetic field is needed, and it is towed at the back of a plane to survey a given area. The instrument is known as the proton precession magnetometer. The major detector element is water held in a nonmagnetic container. A small magnetic field (~100 Oe) is applied momentarily through a solenoid wrapped around the container to align a small fraction of the protons in water and then it is switched off. The only field left is the Earth's magnetic field, which exerts a torque on the magnetic moment of the protons, resulting in their precession and emission of a radio frequency signal. The same solenoid is used to detect this electrical signal. The frequency of this signal ν is related to the magnetic field H by:

$$H(\gamma) = 23.4874\nu(Hz), \tag{9.13}$$

where 1 Oe = $10^5 \gamma$.

9.9.2 MAGNETIC SAMPLING

In some applications, we would like to measure the amount of magnetic materials in a given sample. This can be done by measuring the hysteresis loop of a known volume of the sample, from which we obtain the *apparent* saturation magnetization. Knowing the saturation magnetization of the

---

* There are four magnets in every MRI system. The first is the main magnet that generates a strong, uniform field. The other three are gradient magnets that provide relatively low field strengths (18–27 mT) that vary within the body to be imaged. These gradient fields are switched on at various times (known as slice selection, frequency, and phase encoding) during the data acquisition process.

target material (e.g., M_{sat} (Fe) = 1.7×10^6 amp/m), we can directly deduce its concentration in the sample. In coal- or natural gas-fired power plants, steam turbine temperatures can be measured by inserting steel samples* and measuring the α/γ ratio; α is the magnetic BCC phase and γ is the nonmagnetic FCC phase. The equilibrium α/γ ratio depends on temperature.

9.9.3 ALTERNATORS

An alternator is a generator typically used in cars and planes to convert energy of mechanical motion into electrical energy — a more sophisticated device than that discussed at the beginning of this chapter. It has two sets of wire coils: rotor and stator coils. The rotor is a rigid iron core (typically Fe + 3% Si) with a few dozen turns of wire coil. It is powered by a battery to produce a magnetic field.** The use of Fe + 3% Si ensures the generation of a strong magnetic field of about 1 T at a moderate excitation current. Rotation of the rotor coil by the engine induces an alternating voltage in the adjacent stator coils via the Faraday effect discussed earlier. Typically, three stator coils are wired to produce a three-phase AC voltage output. Using solid-state diodes and a voltage regulator, the AC output is rectified to produce a stable DC voltage that can be used to power various electronics and lighting demands of the car or plane.

Alternators can be contrasted with generators in which the stator carries the magnetic field and the rotor produces the AC voltage. In this case, the output voltage is transferred from the rotor to the outside world with carbon or wire brushes. Because of centrifugal forces, there is concern that brushes could be spun off, causing the generator to explode. As a result, the rotor is normally geared down by a factor of two or more from the engine rotation; such generators typically do not produce significant electrical power until the engine speed exceeds about 1200 revolutions per minute.

9.10 MAGNETIC AND FORCE SHIELDS

9.10.1 MAGNETIC SHIELDS

While the title of this section appears to come from science fiction, the subject matter is real. There are occasions when we want to work in an environment with very low magnetic fields (e.g., measuring the kinetic energy of electrons in certain spectroscopic studies, sensing magnetic fields generated by the brain, etc.). The Earth's magnetic field (0.3–0.5 Oe) has to be shielded. This can be accomplished by surrounding the measurement region with materials having high magnetic permeability. A high-permeability material acts like a sponge for magnetic fields or flux lines, as shown schematically in Figure 9.13.

Consider a two-dimensional external field of B_o interacting with a ring of permeability μ as shown in Figure 9.13. The field inside the ring B is uniform and in the same direction as B_o, given by:

$$B = B_o \frac{9\mu}{(2\mu+1)(\mu+2) - 2\dfrac{a^3}{b^3}(\mu-1)^2}, \tag{9.14a}$$

---

* Such steel samples are not the usual Fe–C alloys, but rather are specially designed to have sufficient oxidation resistance to survive at 600–800°C in these turbines.

** Note that a magnetic field excited by the rotor coil is needed to produce electrical power in an alternator. Therefore, a working battery is necessary for the alternator to function.

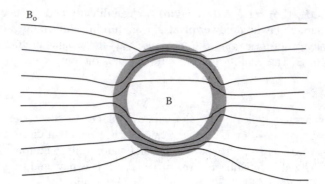

FIGURE 9.13 Magnetic field in the presence of a high-permeability material; B_o = magnetic field outside the ring; B = magnetic field inside the ring.

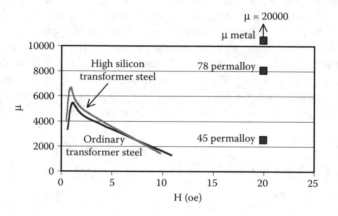

FIGURE 9.14 Permeability versus magnetic field for several selected magnetic alloys. The composition of μ metal is 18Fe 75Ni 2Cr 5Cu (in weight percent). The composition of 78 Permalloy is 21.2Fe 78.5Ni 0.3Mn. The composition of 45 Permalloy is 54.7Fe 45Ni 0.3Mn.

where a is the inner radius of the ring and b the outer radius. In the limit when $\mu \gg 1$, the field inside the ring can be approximated by:

$$B = B_o \frac{9}{2\mu\left(1 - \dfrac{a^3}{b^3}\right)}. \tag{9.14b}$$

Note that the field inside the ring is reduced by a factor of μ. Figure 9.14 shows the permeability values of selected magnetic alloys measured at different magnetic fields.

EXAMPLE

To show the effectiveness of a magnetic shield, consider a ring with an inner radius of 10.0 cm and an outer radius of 10.1 cm (i.e., the magnetic shield is only 1 mm thick). Its permeability is 10^4 in the presence of an external field of 50 μT. Calculate the magnetic field inside.

Solution

Substituting $B_o = 50$, $\mu = 10^4$, $a = 10.0$ cm, and $b = 10.1$ cm into Equation (9.14b), we have:

$$B = 50 \frac{9}{2 \times 10^4 \left(1 - \dfrac{10^3}{10.1^3}\right)} \approx 0.77.$$

Therefore, the magnetic shielding provides a reduction factor of about 65.

9.10.2 FORCE SHIELDS

Consider again the situation of a conductor moving at speed v inside a magnetic field with flux density B. We already know that an induced current I flows inside the conductor. This current is given by:

$$I = \frac{Bwv}{R}, \tag{9.15}$$

where w is the width of the conductor intercepting the magnetic flux lines and R the electrical resistance. The interaction between the induced current and the magnetic field produces a retarding force F, which is given by:

$$F = BwI = \frac{B^2 w^2 v}{R}. \tag{9.16}$$

Note that the retarding force is proportional to the square of the magnetic induction and inversely proportional to the resistance.

EXAMPLE

Consider a metal cube measuring 1 cm on the side moving through a magnetic field, perpendicular to the flux lines, at a speed of 10^3 m/s. The electrical resistivity of the metal is 1×10^{-6} ohm-cm. The magnetic field is 1 T. The density of the metal is 10 g/cm^3. Calculate the instantaneous deceleration experienced by the metal cube.

Solution

The first step is to calculate the retarding force using Equation (9.16), in which $B = 1$ T, $w = 0.01$ m, and $v = 10^3$ m/s. The electrical resistance is given by ρ/w, or 10^{-6} ohm. Therefore, the retarding force F is given by:

$$F = \frac{1^2 (0.01)^2 10^3}{10^{-6}} = 10^5 N.$$

The volume of the metal cube is equal to 1 cm^3, so the mass is equal to 10 g, or 0.01 kg. The deceleration is equal to F/m, or 10^7 m/s^2, or about 1 million g. Therefore, a strong magnetic field can act as a force shield retarding the motion of metal objects. The operation of certain amusement park rides to provide timely retardation is based on this principle.

PROBLEMS

1. Explain why hard magnets are mechanically stronger than soft magnets.
2. Nickel ferrite ($NiFe_2O_4$), a cubic ferrite once used in magnetic read–write heads, has 8 Ni^{2+} and 16 Fe^{3+} ions per unit cell. All 8 Ni^{2+} ions sit in octahedral sites, each with a magnetic moment of $2\mu_B$. Half of the Fe^{3+} ions sit in octahedral sites and half in tetrahedral sites with antiparallel spins. As a result, the Fe^{3+} ions do not contribute to the overall magnetic moment of nickel ferrite. The unit cell side length is 0.834 nm. Calculate the saturation magnetization in amps per meter.
3. A solenoid coil is located inside a uniform magnetic field with a flux density of 2 T. The coil area is 100 cm². The solenoid coil has 50 turns. It is rotated at the rate of 1800 revolutions per minute as shown in Figure 9.15. Calculate the maximum induced voltage appearing across two ends of the solenoid coil.
4. You are given the following magnetic media properties: $M_r = 4 \times 10^5$ amp/m, $H_c = 2 \times 10^5$ amp/m, and the media thickness = 10 nm. At a given temperature, the domain reversal frequency f is given by:

$$f = f_o \exp\left(\frac{\Delta E}{k_B T}\right),$$

where $f_o = 10^9$ s^{-1}, and ΔE is the activation energy, k_B the Boltzmann constant, and T the temperature. In order to ignore thermal effects in high-density disk drives, we require f to be less than 1×10^{-20} per second.
   a. Calculate $\Delta E / k_B T$.
   b. At 300 K, determine the maximum storage density in terabits per square inch as limited by the thermal stability of magnetic domains.
5. a. It is known that each iron atom contributes 2.2 Bohr magnetons at room temperature. Calculate the saturation magnetization of pure iron. There are 8.4×10^{28} iron atoms per cubic meter.
   b. A hysteresis curve is obtained from an iron-containing sample at room temperature. After subtracting the contribution from the external field, $B_{sat} = 1.2 \times 10^{-4}$ T. Calculate the amount of iron in this sample.
6. Certain materials have only one easy magnetization direction. For example, the magnetic domains of cobalt are aligned along the [0001] or [000$\bar{1}$] direction (the c-axis). Argue why there is little or no magnetostriction in this class of magnetic materials.

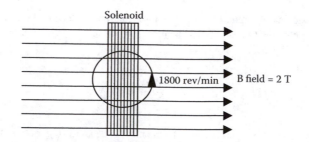

FIGURE 9.15 A solenoid rotating in a magnetic field.

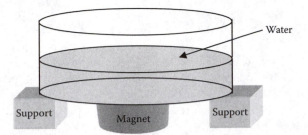

FIGURE 9.16 Demonstrating the diamagnetic property of water.

7. Obtain a petri dish or small dinner plate and fill it with water. Use whatever is available to support the dish or plate so that it is about an inch off the table, as shown in Figure 9.16. Push a strong magnet (cobalt–samarium or neodymium–iron–cobalt) under the center of the plate, as close to the plate as possible. Observe what happens to the water surface. Explain this observation in terms of the diamagnetic response of water.

8. Consider the force shield problem described in the text. With a retarding force equal to αv, where $\alpha = B^2 w^2 / R$ as given in Equation (9.16), the equation of motion is:

$$m \frac{dv}{dt} = -\alpha v,$$

where m is the mass of the object. The negative sign indicates that the force acts in the direction opposite to that of the velocity. Solve this equation to obtain the stopping distance.

10 Thin Films

10.1 WHY THIN FILMS?

In 1956 when hard-disk drives were used to store information in mainframe computers, it took fifty 24-in. disks to store 5 Mbytes (Figure 10.1). This represents an areal storage density of about 2 kbits/in^2. In 2006, the areal storage density is about 100 Gbits/in.^{2}, corresponding to a 50 million-fold increase over 50 years. There are several technologies that make this possible: the use of cobalt-based thin films as the magnetic storage media (instead of γ-Fe$_2$O$_3$ particulates in 1956), the adoption of sensitive thin-film magnetoresistive read heads (instead of inductive coils) and the use of ultrathin protective overcoats that allow the read–write head to fly close to the disk surface for improved sensitivity. In many ways, disk-drive storage represents one of the best examples of how thin films can markedly improve performance and drastically reduce cost.

Another notable example where thin films are critical is the semiconductor industry, in which various thin films are used, from gate oxides in metal-oxide semiconductor devices to diffusion barriers and interconnects between different circuit components. In mechanical components subjected to sliding or rolling (e.g., gears, bearings, and cutting tools), protective coatings provide the last line of defense against friction and wear. With proper control of film composition and architecture, we can fabricate thin-film structures to maximize optical reflectivity, or to suppress it. In this chapter, we will present an overview of how thin films are synthesized and discuss selected properties and applications of thin films.

10.2 DEPOSITION OF THIN FILMS

Generally, we can classify thin-film deposition techniques into two broad categories: physical vapor deposition (PVD) and chemical vapor deposition (CVD). In PVD, materials comprising the thin film are delivered to the substrate by physical means, such as evaporation, sputtering, or laser ablation. In CVD, the thin film is formed on the substrate by a chemical reaction. The demarcation line between PVD and CVD is not distinct. For example, in the process of reactive sputtering of TiN, Ti is sputtered from a target in a background pressure of nitrogen. On the substrate, the arriving Ti atoms react with adsorbed nitrogen to form TiN. In this case, Ti atoms are delivered to the substrate by physical means (sputtering), but the final product TiN is formed by a chemical reaction between Ti and atomic nitrogen. In the molecular beam epitaxy (MBE) synthesis of gallium arsenside, gallium and arsenic vapors produced by sublimation (a physical process) react to form the compound on the substrate. In this section, we present an overview of three deposition techniques: evaporation, sputtering, and chemical vapor deposition.

10.2.1 EVAPORATION

10.2.1.1 Maximum Evaporation Rate and Vapor Pressure

At any temperature above absolute zero, there is some probability that atoms in a solid or liquid may attain sufficient thermal energy to escape into the vapor phase. This probability increases with temperature. As a result, the vapor pressure increases with temperature. Figure 10.2 shows the variation of vapor pressure as a function of temperature for selected elements. In a closed system

50 24" disks; 5 MB

FIGURE 10.1 The IBM 350 hard drive (1956), the primary data storage system for the IBM 305 RAMAC™ system. (Reprinted with permission from IBM Corporation.)

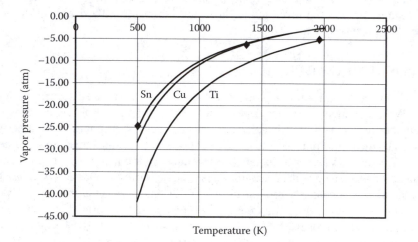

FIGURE 10.2 Variation of $\log_{10}$(equilibrium vapor pressure) versus absolute temperature for tin, copper, and titanium. (♦) marks the melting temperature. 1 atm = 760 torr = 1.01×10^5 Pa.

at thermal equilibrium, the maximum evaporation rate must be equal to that needed to maintain this vapor pressure at the given temperature and is given by:

$$R = \frac{p}{\sqrt{2\pi m k_B T}},$$ (10.1a)

where

R = the maximum evaporation flux (number of atoms leaving unit area of the surface per unit time)

p = the equilibrium vapor pressure

m = the atomic weight

k_B = the Boltzmann constant

T = the absolute temperature

Expressing these variables in practical units, we can write:

$$R = \frac{3.52 \times 10^{22}\, p(torr)}{\sqrt{m(a.m.u.)T(K)}} \Big/ cm^2\text{-s.} \qquad (10.1b)$$

Here, the pressure p is expressed as torr, mass m as atomic mass unit, and temperature T as absolute kelvin.

EXAMPLE

At 1200 K, the vapor pressure of copper is about 10^{-5} torr. Calculate the maximum evaporation flux.

Solution

$p = 10^{-5}$ torr, $m = 63.5$, and $T = 1200$ K. Substituting these numbers into Equation (10.1b), we have:

$$R = \frac{3.52 \times 10^{22} \times 10^{-5}}{\sqrt{63.5 \times 1200}}.$$

Solving, we obtain the maximum evaporation rate of 1.27×10^{15} atoms/cm²-s. Since there are about 1 $\times 10^{15}$ atoms/cm² for a typical solid surface, this evaporation rate is roughly equal to one atomic layer per second.

10.2.1.2 Evaporation Sources

Evaporation is typically done in two ways: resistive heating or electron-beam heating. In resistive heating, the evaporant is loaded onto a refractory container in different forms — for example, open filament, basket, or boat (Figure 10.3) made of tungsten, molybdenum, or tantalum. An electrical current passes through and heats the container. In some cases, to avoid reactivity between the evaporant and the heating element, the evaporant may be contained in an inert crucible made of alumina or graphite indirectly heated by a tungsten or nichrome (Ni–Cr) filament (Figure 10.3).

Another method of heating is electron-beam bombardment. In its simplest configuration, electrons are extracted from a hot tungsten filament and accelerated toward the target (evaporant), which sits on a water-cooled hearth. To eliminate contamination from the tungsten filament, the electrode configuration is such that there is no direct line of sight between the filament and the target (Figure 10.4). Alternatively, a magnetic field is incorporated to steer the electron beam. The advantage of electron-beam heating over resistive heating is that it minimizes contamination and can evaporate virtually any materials.

With negligible contamination from the filament or crucible materials, the cleanliness of evaporated films depends on the purity of source materials and the deposition pressure. For the highest purity films, it is best to deposit thin films under ultrahigh vacuum conditions — that is, pressure better than 1×10^{-9} torr ($\approx 1.3 \times 10^{-7}$ Pa).

10.2.1.3 Evaporation of Alloys

If a binary alloy A_xB_y is being evaporated at a given temperature, there is no assurance that the deposition rate is in direct proportion to the composition of the solid phase. The primary reason is

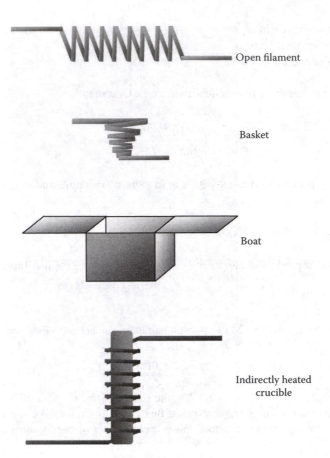

FIGURE 10.3 Different resistively heated evaporation sources.

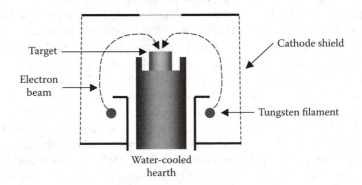

FIGURE 10.4 Schematic setup for an electron-beam evaporator.

that the evaporation rate depends on the vapor pressure of the element. The temperature dependence of vapor pressure may be different for the two elements. A secondary reason is that the vapor pressure of one element may be affected by the presence of another element, due to their mutual interaction.

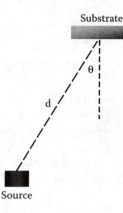

FIGURE 10.5 Deposition geometry showing the relative positions of deposition source and substrate.

10.2.1.4 Dependence of Deposition Rate on Source–Substrate Distance

Assume that the evaporation source is a point source. The deposition rate R_{dep} changes with distance and geometry (Figure 10.5), as given by:

$$R_{dep} \propto \frac{\cos \theta}{d^2}, \tag{10.2}$$

where d is the distance between the source and the substrate and θ the angle between the substrate normal and the direction of the incoming vapor flux.

EXAMPLE

Consider a substrate positioned at 10 cm above a point source. Determine the thickness variation from the center of the substrate (which is immediately above the point source) and a distance of 2 cm from the center of the substrate.

Solution

The deposition rate R_1 at 10 cm above the point source is given by:

$$R_1 = C \frac{1}{(10)^2},$$

where C is some constant. The deposition rate R_2 at 2 cm from the center of the substrate is given by:

$$R_2 = C \frac{1}{(10)^2 + (2)^2} \cos \theta,$$

where

$$\cos \theta = \frac{10}{\sqrt{(10)^2 + 2^2}},$$

as shown in Figure 10.5.

Solving yields:

$$\frac{R_2}{R_1} = \frac{1}{104}\frac{10}{\sqrt{104}}100$$

$$= 0.94.$$

Therefore, at 2 cm from the center, the deposition rate is equal to 94% of that at the center of the substrate.

Question for discussion: Decreasing the source–substrate distance d increases the deposition rate, which is an advantage. What is the disadvantage of doing so? (Hint: Refer to the preceding example.)

10.2.1.5 Deposition Rate Monitors

There are two methods to measure the deposition rate in real time. Both are applicable to other physical vapor deposition techniques such as sputtering:

- Ionization gauge. Within the ionization gauge is a hot filament from which electrons are extracted to ionize evaporated atoms passing through the ionization gauge (Figure 10.6). A collector electrode collects and measures the resulting ion current, which is proportional to the evaporation flux. With proper calibration, the ion current can be related directly to the rate of evaporation.
- Quartz crystal monitor. When atoms are deposited onto a quartz crystal, its mass increases slightly. The resonant frequency of a quartz crystal is inversely proportional to $\sqrt{mass}$. Measurement of the change in resonant frequency Δf gives directly the mass of materials deposited on the quartz crystal, as given by:

$$\frac{\Delta m}{m} = -\frac{\Delta f}{f}, \tag{10.3}$$

where m is the total mass (quartz crystal + evaporant) of the quartz crystal and f its resonant frequency. The minus sign indicates that the frequency decreases with increasing mass. Typical quartz crystal monitors readily detect mass change in the nanogram range.

EXAMPLE

A piece of quartz crystal weighing 0.1 g is designed to have a resonant frequency of 10 MHz. What is the mass sensitivity of such a device if a frequency change of 0.1 Hz can be detected?

Solution

From Equation (10.3), we substitute $m = 0.1$ g, $\Delta f = 0.1$, and $f = 10^7$. Therefore, Δm is given by:

$$\Delta m = \frac{0.1}{10^7} \times 0.1 = 10^{-9} \text{ g.}$$

For a typical metal such as titanium (atomic weight = 48), one atom weighs 8×10^{-23} g. Therefore, 1 ng contains 1.25×10^{13} atoms. When this is deposited onto an area of 1 cm², this is equivalent to about 1% of an atomic layer.

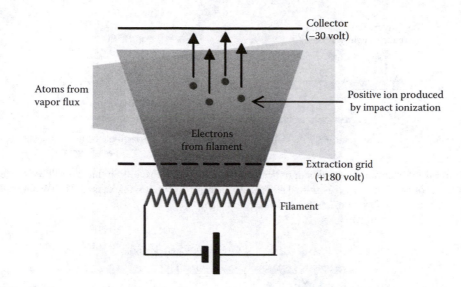

FIGURE 10.6 Schematic diagram of an ionization gauge. The filament is heated to produce electrons, which are extracted by a grid biased at typically +180 V relative to the filament. These electrons collide with atoms in the evaporation flux to produce positive ions, which are collected by a collector electrode biased at typically −30 V relative to the filament. The collector current is proportional to the rate of evaporation.

10.2.1.6 Measurement of Film Thickness

Both techniques described in the preceding paragraph do not give film thickness directly and require some form of absolute thickness calibration. For example, a quartz crystal monitor gives only the weight change deposited on the quartz crystal of a certain area (grams per square centimeter). To convert this information into film thickness requires knowledge of the film density. Given the growth morphology of thin films, film density is not always the same as the density of the corresponding bulk material. Therefore, we need some way to measure the absolute film thickness. Two such methods are applicable to films synthesized by any deposition techniques. The only requirement is the existence of a film step on the substrate surface.

- Diamond stylus profilometer. This method involves loading a diamond stylus lightly on the sample and scanning it across the film as shown in Figure 10.7(a). As the diamond stylus scans across the step produced by the film, its vertical motion is detected by a sensitive displacement transducer, providing an absolute measurement of film thickness. The sensitivity is typically around 0.1–1 nm.

Question for discussion: Why is light loading necessary when scanning the diamond stylus across the film surface to measure film thickness? How will heavier loading affect film thickness measurements?

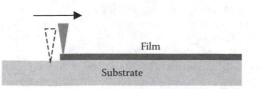

FIGURE 10.7(a) Illustration of diamond stylus profilometry to measure film thickness.

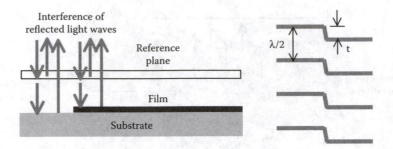

FIGURE 10.7(b) Schematic illustration of optical interferometry to measure film thickness. The interference pattern consists of two sets of fringes shifted from each other due to the film thickness t. The spacing between fringes corresponds to $\lambda/2$, where λ is the wavelength of light.

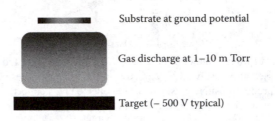

FIGURE 10.8 A typical sputter deposition setup.

- Optical interference. In this method, we place a semitransparent reference plane on the sample as shown in Figure 10.7(b) and then shine monochromatic light (light with a fixed wavelength) onto the sample. Light waves reflected from the back side of the reference plane (which is lightly coated with silver to enhance reflectivity) and the sample surface produce interference fringes as shown on the right side of Figure 10.7(b). The spacing between fringes corresponds to $\lambda/2$, where λ is the wavelength of light. The fringe shift corresponds to t, the film thickness. The sensitivity is typically around 0.1 nm.

10.2.2 SPUTTERING

Evaporation occurs by supplying thermal energy to atoms in the evaporant. Alternatively, we can bombard the target surface with energetic ions, transferring momentum to the atoms and removing them from the target. This process is known as sputtering. Therefore, thin-film deposition can be performed by sputtering without deliberately heating the target.*

In an actual sputter deposition setup, the energetic ions are provided by gas discharge. The sputter gas environment depends on the materials to be deposited. For example, in the sputter deposition of titanium from a titanium target, the sputter gas is argon; in the deposition of titanium nitride from a titanium target, the sputter gas is argon with some small partial pressure of nitrogen. Sputter deposition is normally performed at a pressure between 1 and 10 mtorr. The voltage applied to the target is typically around –500 V (Figure 10.8).

10.2.2.1 Magnetron Sputtering

To increase the deposition rate, the plasma discharge is intensified by incorporating magnets around the target. One typical setup is shown in Figure 10.9(a), which shows top and side views of the

---

* In actual practice, ion bombardment heats up the target. Therefore, the target is water cooling during sputter deposition.

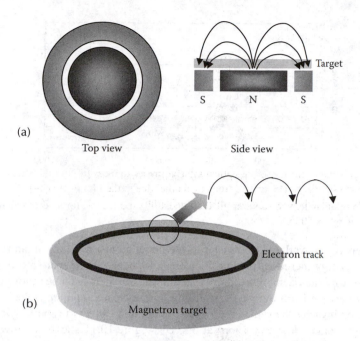

(a)

Top view   Side view

(b)

Magnetron target

FIGURE 10.9 (a) Configuration of the magnetron sputter target; (b) Oscillatory motion of electrons around the magnetron sputter target.

target. Two magnets are behind the target so that a strong magnetic field (0.2–0.5 T) within a few millimeters of the target surface is established. Electrons are trapped by the magnetic flux lines and execute oscillatory motion around the target, as shown in Figure 10.9(b). This increases the probability of ionization and hence the plasma density (i.e., electron and positive ion concentration) in the vicinity of the sputter target. The ions then accelerate towards the negatively biased target. The ion bombardment energy is approximately equal to the target potential. For example, if the target potential is −500 V, the ion bombardment energy is about 500 eV per singly charged ion. This is more than sufficient to remove atoms from the target.

The total number of atoms sputtered from the target is given by two quantities: ion current and sputter yield. The ion current gives the number of ions striking the surface per second. For example, an ion current of 100 mA is equivalent to $0.1/(1.6 \times 10^{-19})$, or about 6.3×10^{17} ions per second (assumed singly charged ions). The sputter yield is defined as the number of atoms removed from the target surface per incident ion. The dependence of sputter yield on ion energy is shown schematically in Figure 10.10. The trend can be understood as follows. As the incident ion energy

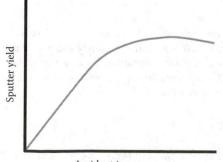

FIGURE 10.10 Schematic plot of the variation of sputter yield versus incident ion energy.

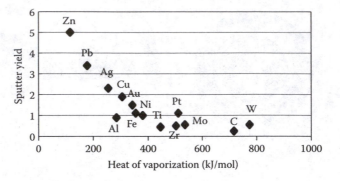

FIGURE 10.11 Correlation between sputter yield and heat of vaporization for selected elements. The argon ion bombardment energy is 500 eV.

increases, more energy is available to transfer to the target atoms, resulting in an increased sputter yield. As the ion energy increases further, incident ions penetrate deeper into the target so that momentum transfer events occur more deeply within the solid. It becomes more difficult for the sputtered atoms to escape from the target, resulting in a decreased sputter yield.

Since sputtering involves the removal of atoms, we should expect the sputter yield to be related to the heat of vaporization. For argon ions at 500 eV, Figure 10.11 shows the inverse correlation between sputter yield and heat of vaporization for selected elements: Higher heat of vaporization results in lower sputter yield.

In typical magnetron sputtering, the ion bombardment energy is less than 1 keV. Under this condition, the sputter yield S is given by:

$$S = \frac{3\alpha}{4\pi^2} \frac{4mM}{(m+M)^2} \frac{E}{U},$$

(10.4)

where

m = the mass of incident ions
M = the mass of target atoms
E = incident ion energy
U = the binding energy of target atoms

The term α is a measure of the energy available for sputtering, depending on the mass ratio and angle of impact; typical range of values is between 0.1 and 1.

EXAMPLE

Consider the sputter deposition of titanium. The conditions of sputtering are as follows: sputter yield of titanium = 0.3, ion current = 100 mA all due to singly charged ions, substrate-to-target distance = 10 cm. Calculate the deposition rate on the substrate. Assume that the sticking probability of sputtered atoms on the substrate is 100% (no reflection or resputtering) and that the sputtered atoms are leaving the target surface in an isotropic distribution.

Solution

With singly charged ions, an ion current of 100 mA is equivalent to $0.1/(1.6 \times 10^{-19})$, or about 6.3×10^{17} ions per second, as discussed earlier. With a sputter yield of 0.3, the number of sputtered atoms = $6.3 \times 10^{17} \times 0.3 = 1.9 \times 10^{17}$ atoms/s.

This atomic flux will be deposited onto a hemisphere with a radius of 10 cm. The total area of this hemisphere is $2\pi(10)^2 \sim 630$ cm^2. Therefore, the deposition rate is given by:

$$\frac{1.9 \times 10^{17}}{630} \approx 3 \times 10^{14} \text{ atoms/cm}^2\text{-s}.$$

EXAMPLE

Calculate the sputter yield of titanium at an argon ion energy of 500 eV. Assume that $\alpha = 0.1$ and $U = 5.0$ eV.

Solution

In this problem, $E = 500$ eV, $m = 39$ a.m.u., and $M = 48$. From Equation (10.3), the sputter yield is equal to:

$$\frac{3 \times 0.1}{4\pi^2} \frac{4 \times 39 \times 48}{(39 + 48)^2} \frac{500}{5.0} = 0.67.$$

The experimental value is in the range of 0.5–0.6.

10.2.2.2 Substrate Bombardment

When atoms are ejected from the target surface, their initial kinetic energies may be as high as the incident ions, typically several hundred electron volts. On their way to the substrate, these atoms may collide with residual gas molecules and hence lose kinetic energies. The average distance between such collisions, known as the mean free path, depends on the residual gas pressure and size and speed of these atoms. For argon atoms at thermal equilibrium with a background pressure p of argon, the mean free path λ is given approximately by:

$$\lambda \cong \frac{5(cm)}{p(mtorr)}, \tag{10.5a}$$

where 1 mtorr = 1×10^{-3} torr.

EXAMPLE

Calculate the mean free path of argon atoms at $p = 5$ mtorr and $p = 5 \times 10^{-6}$ torr.

Solution

Substituting $p = 5$ mtorr into Equation (10.5a), we have:

$$\lambda \cong \frac{5(cm)}{5} = 1 \text{ cm}.$$

For $p = 5 \times 10^{-6}$ torr, we need to convert to millitorrs before we can use Equation (10.5a):

$$p = \frac{5 \times 10^{-6}}{1 \times 10^{-3}} = 5 \times 10^{-3} \text{ mtorr}.$$

Substituting this value of p into Equation (10.5a), we have:

$$\lambda = \frac{5}{5 \times 10^{-3}} = 1000 \text{ cm.}$$

The first case is typical of sputtering. In the second case, the mean free path is so large that molecules collide with the chamber walls more frequently than with other molecules.

A Closer Look

Equation (10.5a) assumes that the atoms of interest have the same average speed and the same size as the background argon atoms. How should we modify that equation if the atom or particle of interest has a different average speed and particle size?

If the particle of interest is moving much faster than the background argon atoms (e.g., an energetic argon atom), the background argon atoms will appear frozen in place as viewed by a fast moving particle. Detailed analysis shows that the mean free path is increased by a factor of $\sqrt{2}$:

$$\lambda \cong \frac{5(cm)}{p(mtorr)} \times \sqrt{2}$$

$$\cong \frac{7(cm)}{p(mtorr)}. \tag{10.5b}$$

If the particle of interest is fast moving and is much smaller than argon (e.g., a fast moving electron), an additional correction factor will be needed. A smaller particle means that the cross section available to collide with the background atoms is smaller. The "collision cross section" for argon–argon collisions is equal to $\pi(2d)^2$, where d is the diameter of an argon atom. For electron–argon collisions, the collision cross section is reduced to πd^2 when we assume the physical cross section of an electron to be near zero. This increases the mean free path by a factor of four:

$$\lambda \cong \frac{28(cm)}{p(mtorr)}. \tag{10.5c}$$

In most deposition systems, the target–substrate separation is often 10 cm or greater. Therefore, a large fraction of sputtered atoms collide with residual gas molecules on their way to the substrate. Consequently, the arrival energies of these sputtered atoms are much lower than their initial energies, typically in the 0.1–10 eV range. Films grown with such low arrival energies tend to have an appreciable density of voids or defects since arriving atoms do not have sufficient mobility to move into ideal lattice sites.

To solve this problem, we can use ions to bombard the surface during film growth. These ions can be provided in two ways. One is to use an auxiliary ion gun to provide energetic ions (typically inert gas ions). Another method is to apply a negative substrate bias to extract ions from the plasma. Optimum ion bombardment of the growing film results in better quality films in terms of reduced defect density and improved crystallinity.

10.2.2.3 Radio Frequency (RF) Sputtering

For conducting materials, sputtering can be performed with direct current (DC) bias voltages (on the target and the substrate). As the electrical resistivity of the target increases, an increasing fraction of the applied voltage drops across the thickness of the target since the applied voltage is the sum of voltage drop across the target and target voltage. At some critical value of electrical resistivity, the voltage drop across the thickness of the target may be so high that the target voltage is insufficient to sustain a plasma discharge unless a large applied voltage is used (exceeding 1000 V). In practice, this critical value is around 10^6 ohm-cm.

EXAMPLE

Consider a target with electrical resistivity of 10^6 ohm-cm. Assuming an active target area of 100 cm^2 and thickness of 0.6 cm, calculate the voltage drop across the thickness of the target at an ion current of 100 mA.

Solution

The resistance of the target = $(10^6 \times 0.6)/100 = 6 \times 10^3$ ohms. At an ion current of 100 mA, the voltage drop = current × voltage = $0.1 \times 6 \times 10^3 = 600$ V. This is a large voltage drop.

The way to get around this problem is to apply radio-frequency (RF) voltage to the target. At sufficiently high frequencies, the impedance of the circuit decreases so that the voltage drop across the target becomes small compared with the applied voltage. Because electrons are lighter and hence more mobile, they respond more quickly to a positive potential than ions responding to a negative potential of the same magnitude. As a result, although the RF voltage applied to a target has a symmetrical waveform (as often positive as it is negative), the target self-biases to a negative potential. Most RF sputter depositions are done at 13.6 MHz.

10.2.3 CHEMICAL VAPOR DEPOSITION

One common example of chemical vapor deposition (CVD) is the deposition of carbon soot films on surfaces from candles, stoves, or barbecue grills. Another example is the reaction between silicon tetrachloride ($SiCl_4$), a volatile compound, and hydrogen to produce silicon films:

$$SiCl_4 + 2H_2 \rightarrow Si + 4HCl. \qquad (10.6)$$

This is an important process used in the electronics industry to deposit single-crystal and polycrystalline silicon films.

In a typical CVD process, a gaseous compound reacts with another gaseous species to produce a solid film on the substrate surface. The process temperature depends on the specific chemical reaction. For example, the reaction shown in Equation (10.6) normally operates at about 1500 K, while the deposition of carbon soot from a candle operates at about 500 K (Figure 10.12). The CVD process can run at low or high pressures. At low pressures, reaction rates can be plasma enhanced (i.e., generation of a plasma discharge of the reactant gases during the deposition process). The reactant gases form excited neutrals, radicals, or charged species during the plasma discharge. These excited species allow film deposition at lower temperatures or at higher rates, or formation of metastable compounds not possible without the plasma. Therefore, this process is known as plasma-enhanced chemical vapor deposition.

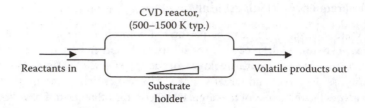

FIGURE 10.12 Schematic diagram of a chemical vapor deposition reactor.

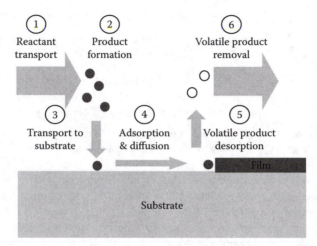

FIGURE 10.13 Illustration of elementary steps in a chemical vapor deposition process.

Several steps are involved in a typical CVD process, as shown in Figure 10.13:

1. Transport of gaseous reactants to the reaction zone
2. Chemical reaction to form product compounds
3. Transport of product compounds to the substrate
4. Adsorption and diffusion of product compounds, followed by film formation
5. Desorption of volatile products
6. Removal of volatile products from the reaction zone

10.2.3.1 Sample Reactions

One critical factor in CVD is the choice of proper chemical reactants (precursors). For example, in the deposition of TiN films using CVD, we need to determine the appropriate chemical precursors for titanium and nitrogen, respectively. Each source must have sufficient vapor pressure to ensure that the deposition rate is not limited by the supply of reactants. In this section, reactions for CVD of selected materials are given.

- Boron carbide ($B_{13}C_3$). Boron carbide is one of the hardest naturally occurring substances at room temperature. Above 1000°C, it is the hardest substance. It is used in abrasives and body armor. The chemical formula is often abbreviated to B_4C. In CVD of boron carbide, the chemical precursor commonly used for boron is diborane (B_2H_6) and that for carbon is methane (CH_4). Using the abbreviated formula, we can write:

$$2B_2H_6 + CH_4 \rightarrow B_4C + 8H_2. \tag{10.7}$$

- Titanium nitride (TiN). TiN is typically used as a wear-protective coating material, as well as for decorative purposes because of its golden color. A common reaction for the CVD of TiN is as follows:

$$2TiCl_4 \text{ (liquid)} + N_2 + 4H_2 \rightarrow 2TiN + 8HCl\uparrow. \tag{10.8}$$

- Tin oxide (SnO_2). Many gas sensors are based on tin oxide thin films. With the introduction of indium, the resulting oxide (indium tin oxide, or ITO for short) is a transparent conductor, used widely in displays. Tin can be provided in the form of organometallic compounds (e.g. tetramethyl tin), as illustrated in the following equation:

$$(CH_3)_4Sn \text{ (liquid)} + 8O_2 \rightarrow SnO_2 + 4CO_2 + 6H_2O. \tag{10.9}$$

Because an organometallic compound is involved, this reaction is sometimes known as organometallic chemical vapor deposition (OMCVD).

- Silicon dioxide (SiO_2). Silicon dioxide is the material that makes microelectronic circuits possible. There are numerous ways to make SiO_2. One is through pyrolysis — that is, thermal decomposition of an organosilicon compound (*pyro* = fire):

$$Si(O\text{-}C_2H_5)_4 \rightarrow SiO_2 + \text{other products (e.g., C, } H_2O, \text{ etc.).} \tag{10.10}$$

- Tungsten. Chemical vapor deposition can also be used to deposit elemental solids. Tungsten is used extensively in microelectronics. Instead of electron-beam evaporation, tungsten is deposited in the microelectronics industry by CVD by reduction of a volatile tungsten compound:

$$WF_6 + 3H_2 \rightarrow W + 6HF\uparrow. \tag{10.11}$$

10.3 STRUCTURE AND MORPHOLOGY

Similar to bulk materials, thin films can be crystalline or amorphous, depending on the growth conditions. Within the crystalline category, films can grow as single crystals or polycrystals. In this case, it is sometimes possible to stabilize a certain crystalline structure by using a certain substrate. For example, tin exists in two crystallographic forms: α-Sn, which has a cubic unit cell and is stable below 13°C, and β-Sn, which has a tetragonal unit cell and is stable above 13°C. When deposited on an indium antimonide (InSb) substrate, which has a cubic unit cell with a dimension almost identical to that of α-Sn, tin acquires the cubic structure at temperatures up to 80°C, well above the normal α–β transformation temperature.

This phenomenon is known as substrate or pseudomorphic stabilization. The driving force for this stabilization is the minimization of free energy associated with the Sn–InSb interface. To minimize elastic strain energy, pseudomorphic growth is more likely to occur if the lattice mismatch between the two film materials is small, preferably less than 15%.

The growth of thin films follows one of three modes, as shown schematically in Figure 10.14:

- Island growth (Volmer–Weber mode). Atoms nucleate as individual islands. As deposition progresses, these islands will eventually coalesce and form a continuous film. This occurs because atoms in the film bond among themselves, rather than with atoms in the substrate.
- Layer-by-layer (Frank–van der Merwe mode). As the name implies, atoms are laid down layer by layer; this occurs when the arriving atoms bond to the substrate. This is an important growth mode for the synthesis of single-crystal films.
- Layer-by-layer initially, followed by island growth (Stranski–Krastanov mode). This is a combination of the first two growth modes. This growth mode occurs due to the presence

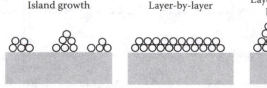

FIGURE 10.14 Three thin-film growth modes.

of strain resulting from a lattice parameter difference between the film and the substrate during the initial layer-by-layer growth (e.g., pseudomorphic growth as discussed in the preceding paragraph). After a few layers, the buildup of strain energy makes the layer-by-layer growth energetically unfavorable. Subsequent deposition results in island growth.

10.4 SELECTED PROPERTIES AND APPLICATIONS

10.4.1 TRANSPORT PROPERTIES

As discussed in the chapter on electrical properties, when the sample dimension such as width or thickness approaches the electron mean free path, properties involving the transport of electrons are affected (e.g., electrical conductivity, thermal conductivity, thermoelectric power, etc.). Consider a thin conducting film with thickness a comparable to the electron mean free path λ. Under this condition, conduction electrons lose energy by an additional mechanism: scattering by the top and bottom surfaces of the thin film. The effect on the electrical conductivity is given by the following equation:

$$\frac{\sigma_f}{\sigma_o} = 1 - \frac{3}{2\kappa}(1-p)\int_1^\infty \left(\frac{1}{t^3} - \frac{1}{t^5}\right)\frac{1 - e^{-\kappa t}}{1 - pe^{-\kappa t}}\, dt, \tag{10.12}$$

where

σ_f = the film conductivity

σ_o = the bulk conductivity

$\kappa = a/\lambda$

a = film thickness

λ = mean free path

p = the fraction of electrons incident on the film surfaces that are specularly or elastically scattered

We can think of p, known as the specularity parameter, as a measure of the fraction of energy lost by conduction electrons upon scattering by the surface. Speculation (never proved) is that p depends on the surface roughness because smoother films give larger p values. In deriving Equation (10.12), we assume that the top and bottom surfaces have the same specularity parameter. Figure 10.15 shows the variation of electrical conductivity as a function of thickness for different specularity parameters.

Note from Figure 10.15 that the electrical conductivity decreases as the film thickness becomes smaller than or on the order of the mean free path. Since electrons are the primary carriers of heat energy in metals, the thermal conductivity of metallic films follows the same dependence on film thickness as that shown in Equation (10.12).*

---

* Heat is also transported by lattice vibrations (phonons). As a result, the effective mean free path for the conduction of heat is not the same as that for the conduction of electrical charge. However, the functional form represented by Equation (10.12) is still valid for the reduced thermal conductivity (the ratio of film thermal conductivity to bulk thermal conductivity).

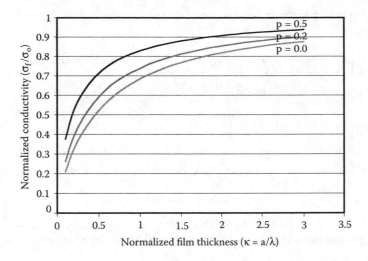

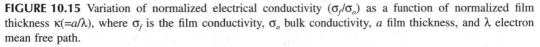

FIGURE 10.15 Variation of normalized electrical conductivity (σ_f/σ_o) as a function of normalized film thickness $\kappa(=a/\lambda)$, where σ_f is the film conductivity, σ_o bulk conductivity, a film thickness, and λ electron mean free path.

10.4.2 OPTICAL PROPERTIES

Optical coatings are applied to surfaces for three different reasons: cosmetic or decorative, suppressed reflectivity, and enhanced reflectivity. We will discuss each one.

10.4.2.1 Cosmetic or Decorative Coatings

One example of a decorative coating is titanium nitride (TiN), which has a golden color and hence serves well for decorative purposes. Zirconium carbide (ZrC) has a battleship-gray color and is often used to coat door hardware, faucet handles, and various household fixtures. TiN and ZrC are hard materials and offer excellent wear protection. As shown in Figure 10.16, titanium dioxide (TiO_2) absorbs ultraviolet (UV) light with a wavelength shorter than 400 nm and is often mixed into paint to minimize discoloration due to UV light-assisted oxidation.*

10.4.2.2 Suppressed Reflectivity

Antireflection coatings are applied to prescription and photographic lenses for enhanced light transmission and reduced glare. The design principle is based on the interference of light reflected from the air–coating and coating–substrate interfaces, as shown in Figure 10.17.

Consider a light wave incident on a surface coated with a thin film, thickness t. For light at normal incidence, when the following two conditions are satisfied, reflected light waves interfere destructively, resulting in no reflected light:

$$2n_c t = \frac{\lambda}{2};$$ (10.13a)

$$n_c = \sqrt{n_a n_s},$$ (10.13b)

* TiO_2 is also incorporated into cosmetics in the form of nanometer-sized particles. The rationale is that TiO_2 absorbs UV light and hence reduces damage to the skin.

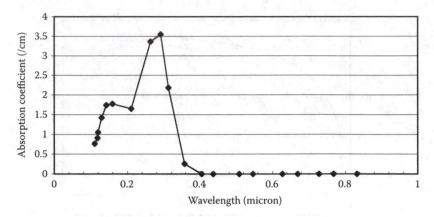

FIGURE 10.16 Experimental data showing the absorption coefficient of TiO_2 as a function of wavelength. The abrupt increase of absorption at a wavelength less than 0.4 µm corresponds to the bandgap energy of 3.0 eV for TiO_2.

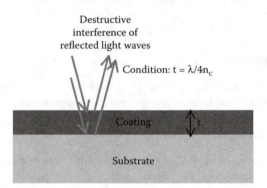

FIGURE 10.17 Antireflection coating by destructive interference. The destructive interference condition is: $t = \lambda/4n_c$, where t is the coating thickness, λ wavelength of light, and n_c refractive index of the coating.

where

t = the coating thickness
λ = the wavelength of light
n_c = the refractive index of the coating
n_a = the refractive index of air
n_s = the refractive index of the substrate

Question for discussion: Equation (10.13a) is the condition for destructive interference; that is, the path difference between light waves reflected from the air–coating interface and the coating–substrate interface is equal to half the wavelength. Why is the second condition as shown in Equation (10.13b) necessary?

EXAMPLE

For a glass substrate (refractive index = 1.5), determine the refractive index and thickness of the coating needed to obtain perfect antireflection at a 500-nm wavelength.

Solution

In this problem, $\lambda = 500$ nm, $n_a = 1$, and $n_s = 1.5$. From equation (10.13b), the refractive index of the coating n_c is given by:

$$n_c = \sqrt{1 \times 1.5} = 1.225.$$

From Equation (10.13a), the optimum coating thickness t is given by:

$$t = \frac{\lambda}{4n_c} = \frac{500}{4 \times 1.225} = 102 \text{ nm.}$$

There is no suitable coating with the correct refractive index for glass substrates. The closest material is magnesium fluoride (MgF_2), which has a refractive index of 1.38 in the middle of the visible spectrum.

Question for discussion: The antireflection conditions shown in Equation (10.13) apply to light with one specific wavelength. Discuss the color appearance of a coating designed to satisfy Equation (10.13) for green light only.

10.4.2.3   Enhanced Reflectivity

During the summer, most of us have experienced the oven-like sensation of getting into a passenger car that has been under sunlight for an hour or longer. This is partly due to the greenhouse effect provided by the windows and partly due to the absorption of the solar radiation by the paint of the car body. For example, a red car absorbs about 50% of the incident solar radiation, increasing to about 70% for the same car with black paint. It would be desirable to develop coatings with enhanced reflectivity of solar radiation for transportation applications such as cars and planes.*

A more common application for coatings with enhanced reflectivity is in mirrors. This is especially important in situations in which the light is reflected many times (e.g., laser mirrors used in optoelectronic research and telecommunications). Near normal incidence, the reflectivity R or fraction of light reflected at an interface between two materials with refractive indices n_1 and n_2 is given by:

$$R = \left(\frac{n_1 - n_2}{n_1 + n_2} \right)^2. \qquad (10.14)$$

Equation (10.14) suggests one method to obtain enhanced reflectivity: depositing a thin film on a substrate so that there is a large difference between the two refractive indices. For example, compare the deposition of alumina (refractive index = 1.8) with titanium dioxide (refractive index = 2.4) onto glass (refractive index = 1.5), all refractive indices given for light in the middle of the visible spectrum (wavelength ~ 500 nm). Straightforward substitution shows that the reflectivity is equal to 0.8% for the alumina–glass interface and 5.3% for the titanium oxide–glass interface.

---

* There is an added bonus if such coatings can be developed. With reduced transmission of ultraviolet light, polymer components within the vehicle are less susceptible to degradation and thus last longer.

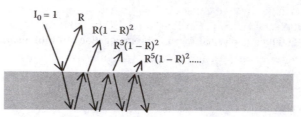

FIGURE 10.18 Reflection and transmission of light. With an incident light intensity of unity, the first reflection intensity is R, the second reflection $R(1-R)^2$, etc. The total reflection intensity is then equal to $R + R(1-R)^2 + R^3(1-R)^2 + R^5(1-R)^2 + \ldots$, which sums to $2R/(1+R)$.

EXAMPLE

Calculate the fraction of light reflected by a glass slide in air. Assume the refractive index of air to be 1.0 and of glass to be 1.5. Assume no light absorption by glass.

Solution

Examine Figure 10.18. Consider the first light ray incident on the front glass surface. The fraction of light R reflected is given by:

$$R = \left(\frac{1-1.5}{1+1.5}\right)^2 = 0.04.$$

Therefore, 96% $(1-R)$ of the light gets into the glass. On the back glass surface, the fraction of original light intensity reflected is equal to $0.96R = 0.0384$. As this light ray exits the front surface, the fraction of the original light intensity transmitted is equal to $0.0384 \times 0.96 = 0.037$. This multiple reflection sequence continues, and the final result for the total reflectivity R_{total} due to multiple reflections is equal to:

$$R_{total} = \frac{2R}{1+R} = \frac{2 \times 0.04}{1.04} = 0.077.$$

Therefore, about 8% of the incident light is reflected by a piece of glass.

The second method to enhance reflectivity is to take advantage of constructive interference of light scattered from multiple thin-film layers. Figure 10.19 shows the stacking of two different thin-

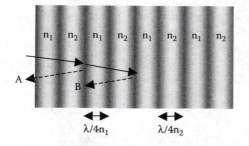

FIGURE 10.19 Coating architecture to provide for enhanced reflectivity; n_1 and n_2 are refractive indices of two thin-film materials; λ = wavelength of light.

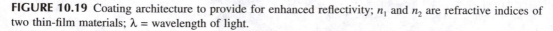

film materials in a periodic arrangement. Each layer has a thickness equal to one quarter the wavelength of light within the material. For light waves scattered from two adjacent periodic stacks (A and B in Figure 10.19), the path difference between them is equal to one wavelength. This results in constructive interference between these two light waves and hence enhanced reflectivity.

EXAMPLE

To obtain enhanced reflectivity for infrared light with wavelength λ equal to 1000 nm with multilayer coatings of titanium dioxide ($n_1 = 3.0$) and silicon dioxide ($n_2 = 1.55$), calculate the individual layer thicknesses.

Solution

For titanium dioxide layers, the thickness of each layer should be equal to $\lambda/4n_1$:

$$\frac{1000}{4 \times 3.0} = 83 \text{ nm.}$$

For silicon dioxide layers, the thickness of each layer should be equal to $\lambda/4n_2$:

$$\frac{1000}{4 \times 1.55} = 161 \text{ nm.}$$

10.4.3 MECHANICAL PROPERTIES

Experimentally, it was found that thin films tend to be stronger than their bulk counterparts. This is reminiscent of the Hall–Petch effect discussed in the chapter on mechanical properties. For polycrystalline films, the average grain size is usually about the same order as the film thickness. As a result, thinner films give smaller grain size and hence higher strength. For example, the yield strength of bulk gold is on the order of tens of MPa, while 200-nm thick gold films have yield strengths in excess of 100 MPa.

Similar to bulk materials, thin films can be made of more than one component. With thin films made of two nanoscale phases, it was discovered that they attain a hardness above the rule-of-mixture values. Such thin films can be in the form of nanocomposites or nanolayers (Figure 10.20). There are several reasons for the hardness enhancement:

- Grain boundaries and interfaces between layers act as obstacles against dislocation motion.
- Dislocations encounter a barrier as they traverse from the phase with lower shear modulus to one with higher shear modulus.
- When the grain size is below about 10 nm, it is energetically unfavorable to generate dislocations. Therefore, such grains are dislocation free, which makes them strong.

10.4.3.1 Hardness

Hardness of thin films is determined by nanoindentation, as discussed in the chapter on mechanical properties. In this technique, a diamond tip is pressed against the thin film of interest. The nanoindentation hardness is defined as:

$$H_{nano} = \frac{P}{A_{proj}}, \qquad (10.15)$$

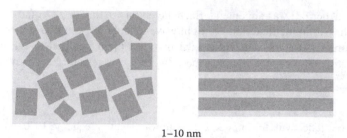

1–10 nm

FIGURE 10.20 Schematic diagram showing the structure of (a) nanocomposite and (b) nanolayer thin films. The two shaded features represent two different phases in the nanocomposite/nanolayer materials.

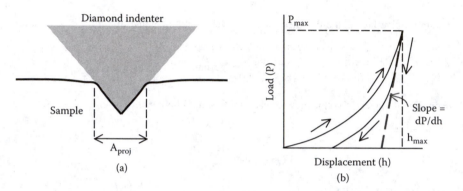

FIGURE 10.21 Measurement of hardness and elastic modulus by nanoindentation: (a) diamond tip indenting into the sample; (b) schematic load-displacement curve obtained in nanoindentation.

where H_{nano} is the nanoindentation hardness, P the normal load, and A_{proj} the area of contact between the diamond tip and the film surface projected along the direction of normal load, as illustrated in Figure 10.21.

Because of the small contact size, the area of contact is not measured directly; rather, it is deduced from the displacement of the tip into the film and the shape of the diamond tip. For an ideal Berkovich diamond indenter commonly used in nanoindentation, the relationship between the projected area of contact and the distance from the apex of the indenter is given by:

$$A_{proj} = 24.5h^2, \tag{10.16}$$

where h is the distance from the apex of the indenter. This is known as an area function. For a real diamond indenter, the area function has to be obtained by calibration or other means, and the projected area of contact at the maximum load is determined by a specific procedure, as discussed in the appendix of this chapter. As a general practice, the penetration distance is typically held to less than 15% of the film thickness. This is to avoid elastic and plastic effects from the substrate.

EXAMPLE

A Berkovich diamond tip indents into a thin film by applying a load of 3 mN. The tip penetration into the film is such that the projected area of contact at 3-mN load is measured to be 2.45×10^{-13} m². Calculate the hardness of this film.

Solution

Since the load is 3×10^{-3} N and $A_{proj} = 2.45 \times 10^{-13}$ m^2, from Equation (10.15), we have:

$$H_{nano} = \frac{3 \times 10^{-3}}{2.45 \times 10^{-13}} = 1.22 \times 10^{10} \text{ Pa}$$

$$= 12.2 \text{ GPa.}$$

Therefore, the hardness of this thin film is 12.2 GPa.

10.4.3.2 Elastic Modulus

In addition to hardness, nanoindentation can be used to determine the elastic modulus of thin films. In this case, we perform a load-displacement excursion on the thin film of interest, as shown in Figure 10.21, loading the film up to a maximum load P_{max} (with maximum displacement h_{max}). This is followed by reducing the load to zero. The slope of the curve upon unloading from P_{max} is known as the stiffness of the contact S, given by:

$$S = \left.\frac{dP}{dh}\right|_{P_{max}} = 2E^* \sqrt{\frac{A_{proj}}{\pi}} , \tag{10.17}$$

where P is the load, h the displacement, and A_{proj} the projected area of contact. E^* is known as the reduced or effective elastic modulus of the contact, given by:

$$\frac{1}{E^*} = \frac{1 - v_d^2}{E_d} + \frac{1 - v_f^2}{E_f} , \tag{10.18}$$

where v and E are the Poisson ratio and elastic modulus, respectively, and the subscripts d and f represent the diamond tip and the film, respectively. For a diamond indenter, $E_d = 1200$ GPa and $v_d = 0.2$ so that the term $(1 - v_d^2)/E_d$ is equal to 0.0008 GPa^{-1}.

EXAMPLE

Consider performing a nanoindentation experiment on a thin film with a diamond indenter. The stiffness of the contact at a certain maximum load is measured to be 10^5 N/m. The projected area of contact at this maximum load is found to be 6×10^{-14} m^2. Assuming a Poisson ratio of 0.3 for the film, calculate its elastic modulus.

Solution

According to Equation (10.17), we can express the effective elastic modulus of the contact E^* as follows:

$$E^* = \frac{S}{2} \sqrt{\frac{\pi}{A_{proj}}} .$$

Since $S = 10^5$ N/m and $A_{proj} = 6 \times 10^{-14}$ m², we can solve for E^*:

$$E^* = \frac{10^5}{2} \sqrt{\frac{\pi}{6 \times 10^{-14}}}$$

$$= 3.62 \times 10^{11} \text{ Pa}$$

$$\approx 362 \text{ GPa}.$$

Substituting this back into Equation (10.18), we have:

$$\frac{1}{362} = 0.0008 + \frac{1 - (0.3)^2}{E_f},$$

which reveals E_f (elastic modulus of the film) to be 464 GPa.

10.4.3.3 Intrinsic Stress

Whether grown by chemical or physical vapor deposition techniques, films often acquire intrinsic stress. The stress arises from a number of sources — for example, the implantation of background gases such as argon during sputtering, bombardment of the growing film by energetic particles (atomic shotpeening), and the difference in thermal expansion between the substrate and the film. The stress can be tensile or compressive, ranging from close to zero to several GPa.

One way to measure the film stress is by the substrate curvature method. When the film stress is compressive (tensile), the film substrate bends concave downward (upward), as shown in Figure 10.22. The Stoney equation gives the relationship between the film-substrate curvature and the film stress:

$$\sigma_f = \frac{E_s t_s^2}{6R(1 - v_s)t_f}, \tag{10.19}$$

where

  σ_f = the film stress
  E_s = elastic modulus of the substrate
  R = radius of curvature of the film-substrate composite
  v_s = the substrate Poisson ratio
  t_f = the film thickness

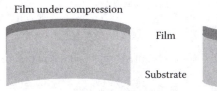

FIGURE 10.22 Curvature of the film-substrate system when the film is under compression and tension.

EXAMPLE

A thin film of titanium nitride with thickness of 100 nm is deposited onto an initially flat silicon substrate with 100-μm thickness. Because of compressive stress in the film, the sample becomes curved (concave downward), with the radius of curvature equal to 10 m. Given that the elastic modulus of silicon is 140 GPa and that its Poisson ratio is 0.25, determine the film stress.

Solution

It is important to keep consistent units. In this problem, film stress (σ_f) and elastic modulus (E_s) will be in GPa, while thicknesses (t_s and t_f) and radius of curvature (R) will be in meters:

$$E_s = 140 \text{ GPa}; \ t_s = 100 \times 10^{-6} \text{ m}.$$

$$R = 10 \text{ m}; \ t_f = 100 \times 10^{-9} \text{ m}.$$

The Poisson ratio for silicon v_s is given as 0.25. Substituting these numbers in Equation (10.19), we can solve for the film stress σ_f as follows:

$$\sigma_f = \frac{140 \times (100 \times 10^{-6})^2}{6 \times 10 \times (1 - 0.25) \times 100 \times 10^{-9}}$$

$$= \frac{1.4 \times 10^{-6}}{4.5 \times 10^{-6}}$$

$$\approx 0.3 \text{ GPa}.$$

The film/substrate sample curvature R in Equation (10.19) can be obtained by optical reflection or, more commonly, by diamond stylus profilometry. In the latter technique, the diamond stylus is scanned a certain horizontal distance across the sample surface and the vertical displacement of the stylus is measured. The curvature R is then given by:

$$R = \frac{L^2}{2z}, \tag{10.20}$$

where L is the horizontal distance scanned and z the vertical displacement of the stylus. Equation (10.20) is valid when $R \gg z$, as is usually the case.

10.4.4 FRICTION AND WEAR PROPERTIES

10.4.4.1 Friction and Wear

Thin films or coatings are often applied to surfaces to reduce friction and wear. Friction performance is measured by the friction coefficient μ, which is defined as the ratio of the friction force to the normal load. Two laws of friction are valid under limited load, speed, and other testing conditions:

- The friction force is proportional to normal load and is independent of the apparent area of contact (Amonton's law).
- Friction is independent of the sliding speed (Coulomb's law).

TABLE 10.1
Selected Coatings and Applications

Coating	Sample Applications
Titanium nitride (TiN)	Tool bits, decorative coating
Cubic boron nitride (c-BN)	Mining drill bits
Titanium aluminum nitride ($Ti_xAl_{1-x}N$)	Cutting and machine tools
Diamond	Cutting and machine tools, drill bits
Diamond-like carbon	Computer disk drives, razor blades
Zirconium carbonitride	Bathroom fixtures and hardware
Molybdenum disulfide	Satellite components
Chromium nitride	Cutting and machine tools

By definition, friction means energy dissipation, which occurs mainly in the form of heat, plastic deformation, fracture, and generation of wear particles. In the United States, it has been remarked that friction results in a 6% loss of the gross domestic product, due to energy inefficiency, wear, and premature failure. For a $10 trillion economy, that is significant. Various coatings have been developed to minimize friction and wear under different service conditions. Table 10.1 gives a sampling of coatings in use today.

Many of today's tribological* components (e.g., bearings, gears, or cutting tools) or systems (e.g., automotive engines) cannot operate without some sort of coating. Under appropriate service conditions, coatings such as diamond-like carbon and molybdenum disulfide can produce ultralow friction coefficients (i.e., less than 0.01), while coatings such as titanium aluminum nitride can provide wear protection for cutting tools without coolants up to 800°C.

EXAMPLE

Consider an object sliding on a surface in a straight line. Its mass is 1000 kg and the coefficient of friction for the system is 0.1. What is the energy lost to friction after sliding through 30 km?

Solution

By definition, the coefficient of friction is given by:

$$\mu = \frac{F}{N},$$

where μ is the coefficient of friction, F friction force, and N the normal load. In this example, $\mu = 0.1$ and $N = 1000\ g = 9.8 \times 10^3$ newton (g is the acceleration due to gravity, 9.8 m/s^2). Therefore, the friction force is given by:

$$F = \mu N = 0.1 \times 9.8 \times 10^3$$

$$= 9.8 \times 10^2 \text{ newton.}$$

The work done against friction after sliding 30 km = F × sliding distance = $9.8 \times 10^2 \times 30 \times 1000$ = 2.94×10^7 J. As a comparison, the energy obtained from the complete combustion of 1 L of gasoline is equal to 3.6×10^7 J.

---

* The study of friction, wear, and lubrication is known as tribology (*tribos* in Greek means "rubbing").

Note that friction and wear are system parameters; that is, they depend on the system as a whole. The system includes the two surfaces in contact, the lubricant used, the surrounding environment, and the testing conditions (load, contact geometry, speed, and temperature).

10.4.4.2 Wear Mechanisms

From an engineering viewpoint, wear is an important performance parameter when two surfaces slide against each other. Four major wear mechanisms have been identified:

- Abrasive wear. When hard particles are present at the sliding interface, they may plow into the surface of interest and produce wear (Figure 10.23).
- Adhesive wear. Wear can be caused by material from the surface of interest adhering to the counterface material. For metallic systems, this is more likely to occur when two similar materials are sliding against each other, as measured by the degree of mutual miscibility (Figure 10.24). For example, bare steel surfaces sliding against each other produce more wear than when one surface is coated with Ni.
- Corrosive or tribochemical wear. In this case, a chemical reaction occurs at the interface resulting in the removal of material. This can be induced by corrosion (when multiple phases are present in humid air), material fracture exposing reactive surfaces, or frictional heating. For example, when sliding experiments are conducted using diamond-like carbon in air, carbon in the film reacts with oxygen in the air to produce carbon dioxide (Figure 10.25), resulting in the thinning of the diamond-like carbon film.
- Fatigue. This type of wear often occurs in bearings subjected to cyclic stresses. Under low-friction conditions on smooth ball bearings, the maximum stress occurs below the surface. As a result, fatigue cracks form beneath the surface (thus, such cracks cannot be detected visually). Eventually, these cracks coalesce and form a wear particle known as fatigue spall (Figure 10.26).

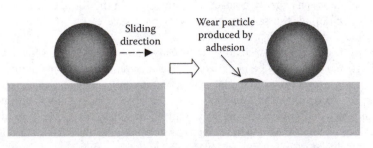

FIGURE 10.23 Schematic illustration of abrasive wear.

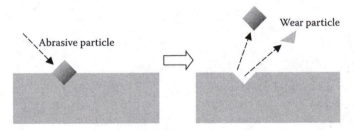

FIGURE 10.24 Schematic illustration of adhesive wear.

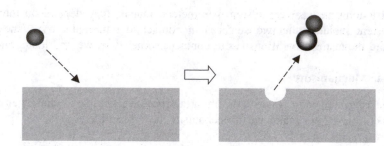

FIGURE 10.25 Schematic illustration of corrosive or tribochemical wear.

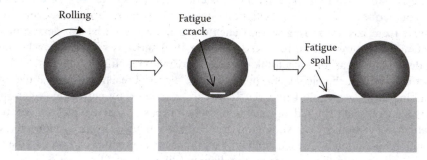

FIGURE 10.26 Schematic illustration of fatigue wear.

10.4.4.3 Archard's Law

Archard's law describes the relationship between wear rate and normal load applied to the contact, as follows:

$$Q = k\frac{N}{H},$$ (10.21)

where

      Q = the wear volume per unit sliding distance
      k = some dimensionless constant
      N = the normal load
      H = hardness of the material

Since surfaces are not absolutely flat and smooth, contacts between surfaces occur at asperities. Such asperity contacts occur over a small fraction of the apparent area of contact. The constant k is the probability of a given asperity contact event resulting in the formation of a wear particle. As a general practice, k defines the severity of the wear rate: severe ($k > 10^{-3}$), moderate (k between 10^{-3} and 10^{-5}), and mild ($k < 10^{-5}$).

As with many friction laws, Archard's law is only valid under limited load, speed, and other testing conditions. For example, the law predicts that the wear rate (wear volume per unit sliding distance) is inversely proportional to hardness. Depending on the mechanism responsible for wear, hard surfaces and coatings do not always guarantee a low wear rate. Under mild wear conditions, the wear rate increases approximately linearly with normal load, in accordance to Archard's law. However, when the contact stress is above approximately one third of the material hardness, wear rate increases sharply, most likely due to severe plastic deformation.

It is useful to compare the wear performance of materials based on the wear rate normalized by the load — that is, wear volume per unit sliding distance per unit load (Q/N), usually expressed as cubic millimeters per newton-meter. According to Archard's law, we can write:

$$\frac{Q}{N} = \frac{k}{H} = k',$$

(10.22)

where k' is known as the wear coefficient. Note the relationship between the wear coefficient and wear probability k.

EXAMPLE

Consider the use of coatings to improve the wear life of a given tribological component. From other experiments, the wear coefficient of this coating material is known to be 5×10^{-11} mm³/N-m under certain testing conditions. Under the same set of testing conditions, with an applied load of 5N and effective contact area of 10^{-4} mm², determine the lifetime of a 10-μm thick coating in meters.

Solution

From Equation (10.22), we can calculate the wear rate at an applied load of 5N:

$$\frac{Q}{5} = 5 \times 10^{-11}.$$

This means that $Q = 2.5 \times 10^{-10}$ mm³/m. Recall that Q is the wear volume per unit sliding distance. The total wear volume available in a coating with a thickness of 10 μm and effective contact area of 10^{-4} mm² is given by:

$$V = (10 \times 10^{-3}) \times 10^{-4}$$

$$= 10^{-6} \text{ mm}^3,$$

where V is the total wear volume available. Therefore, the lifetime of the coating L is given by:

$$L = \frac{V}{Q} = \frac{10^{-6}}{2.5 \times 10^{-10}}$$

$$= 4000 \text{ m}.$$

Therefore, the lifetime is 4000 m.

10.4.4.4 Wear Rate and Plasticity Index

Since multiple mechanisms are responsible for wear, it is not possible to derive a single equation or parameter that can tell an engineer which coatings to use to obtain the lowest wear rate under all conditions. For example, when abrasive wear dominates, coating hardness is a good figure of merit. When corrosive or tribochemical wear dominates, reactivity of the components for a specific chemical reaction becomes important. When plastic deformation is the controlling factor, there is a relatively simple parameter to rank coatings: the H/E ratio. In the study of contacts between asperities, the onset of plastic deformation for asperities is less likely to occur when a certain parameter Ω is as small as possible, defined by:

$$\Omega = \frac{E}{H}\left(\frac{\sigma}{R}\right)^{1/2},$$ (10.23)

where

E = the elastic modulus
H = the hardness
σ = root-mean-square surface roughness
R = the average radius of curvature of asperities

The parameter Ω is known as the plasticity index. Therefore, to minimize plastic deformation (and hence wear resulting from plastic deformation), we should use coatings and surfaces with large H/E ratios.

The value of H/E for coatings can be obtained by analyzing nanoindentation data in two ways. One has been described earlier in this section. The other is through the idea of plastic work during nanoindentation. Figure 10.27 shows the total work and plastic work done during the cycle of loading and unloading in a nanoindentation experiment. Theoretical analysis shows the following relationship to be valid under a wide range of conditions:

$$\frac{H}{E^*} = 0.2 - 0.2\frac{W_p}{W_{tot}},$$ (10.24)

where E^* is the effective elastic modulus for the tip–surface contact as defined by Equation (10.18), W_p the plastic work, and W_{tot} the total work done during the nanoindentation experiment, as shown in Figure 10.27. Therefore, measurement of W_p/W_{tot} gives directly H/E^* (and hence H/E). For example, a soft metal such as aluminum gives H/E^* around 0.003, while a harder metal such as tungsten gives H/E^* of 0.014. As a comparison, the H/E^* value is about 0.05 for titanium nitride and 0.13 for certain diamond-like carbon films.

Question for discussion: Assume that H and E^* (or E) can be independently controlled. One method to increase the H/E^* ratio is to reduce E^*. What is the disadvantage of having small elastic

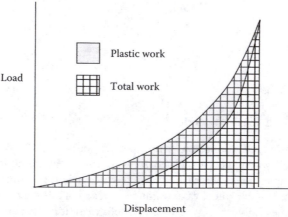

FIGURE 10.27 Illustration of total work and plastic work in a nanoindentation experiment.

modulus values for heavily loaded wear-protective coatings? (Hint: How would small E^* values affect the stiffness of the contact? Will such a stiffness change adversely affect the wear-protection function of the coatings?)

10.5 BIOMEDICAL APPLICATIONS

There have been numerous exploratory studies on the use of coatings to provide improved wear and corrosion performance for implant materials used in joints and vascular stents. In order for such coatings to be adopted for medical use, they must not be toxic to the human body or induce any immune response. For blood-contacting biomedical devices, it is critical that such coatings will not cause the adsorption of blood cells and formation of blood clots. So far, diamond-like carbon coatings appear to provide the best performance. Given our current ability to synthesize various nanocomposite/nanolayer coatings with different chemical modifications, the variety of coatings and the type of biomedical applications will only increase with time. It could even be argued that coatings will be an integral part of an improved quality of life for millions to come.

APPENDIX: OBTAINING THE PROJECTED AREA OF CONTACT IN NANOINDENTATION EXPERIMENTS

To obtain the area of contact in a nanoindentation experiment is more than just getting the maximum penetration depth h_{max} at the maximum load P_{max}. As the tip retracts during unloading, the tip–surface contact is not immediately lost due to elastic "rebound" of the surface. Therefore, the area of contact corresponds to an indenter penetration depth h_c (contact depth) somewhat less than h_{max}. Following Oliver and Pharr and referring to Figure 10.28, we can write the following equation to obtain the contact depth h_c:

$$h_c = h_o + 0.25(h_{max} - h_o), \qquad (10.25)$$

where h_o is the x-axis intercept from the extrapolation of the tangent to the load-displacement curve at the maximum load P_{max}. The projected area of contact is equal to the tip area function evaluated at h_c — that is, $A_{proj}(h_c)$.

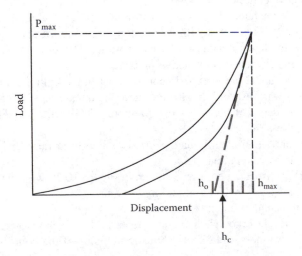

FIGURE 10.28 Determination of the contact depth h_c from nanoindentation load-displacement data. Note that h_c lies one quarter of the way between h_o and h_{max}, as shown.

The tip area function is usually obtained by performing indentation experiments at different loads on a standard material with a known value of $E/(1 - v^2)$, where E is the elastic modulus and v Poisson ratio of the standard material. Fused quartz is a commonly used standard, with $E/(1 - v^2)$ equal to 69.9 GPa.

PROBLEMS

1. Consider the sputter deposition of titanium using argon with the following parameters: target ion current = 300 mA; sputter yield = 0.3; target-substrate distance = 10 cm.
   a. Assuming that the target ion current is due to singly charged ions and that titanium atoms stick onto the substrate with 100% sticking probability, calculate the deposition rate on the substrate.
   b. The deposition chamber has a background pressure of water vapor of 1×10^{-8} torr. This means that the film surface will be bombarded by water molecules at a certain rate. At 300 K, what is the bombardment rate in number of molecules per square centimeter per second?
   c. From these results, what is the water impurity content in the titanium film, assuming 100% sticking probability for titanium and for water?
2. The following data are obtained for the reduced electrical conductivity of copper films prepared by evaporation versus reduced thickness:

Reduced Thickness	Reduced Electrical Conductivity
0.25	0.55
0.50	0.65
0.75	0.74
1.0	0.78
1.50	0.84

   Assume that Equation (10.12) is applicable to this situation. Based on these results, determine the specularity parameter p. (Hint: You need to evaluate Equation 10.12 for various specularity parameter values.)
3. In order to provide enhanced reflectivity for light with a wavelength equal to 600 nm, a multilayer approach is adopted with two materials having refractive indices of 1.5 and 3.5. Determine the layer thickness of each material. Discuss the advantage of having large differences between two refractive indices.
4. In performing nanoindentation experiments on a 1-μm thick titanium nitride film deposited on silicon, it was observed that the measured hardness decreased from about 25 GPa at an indentation depth of about 0.1 μm to about 14 GPa at an indentation depth of 2 μm. Explain these observations.
5. Discuss conditions when a hard coating provides low wear rates and when it does not. Give an example for each situation.
6. Consider the load-displacement curve shown in Figure 10.28 with the following parameters: P_{max} = 2 mN; h_{max} = 100 nm; and h_o = 80 nm.
   a. Determine the contact depth (h_c).
   b. Assuming an ideal Berkovich diamond indenter, calculate the projected area of contact.
   c. Calculate the hardness and effective elastic modulus of the contact.

7. A sliding wear experiment is performed on a given protective coating at an applied load of 0.5 N. The total wear volume after sliding 100 m is equal to 10^{-3} mm^3. The coating hardness is 20 GPa.

 a. Determine the wear coefficient. Express the units in cubic meters per newton-meter.

 b. Assuming the validity of Archard's law, determine whether the wear rate is mild, moderate, or severe.

Bibliography

GENERAL TEXTBOOKS

Ashby, M., and K. Johnson. 2002. *Materials and Design, the Art and Science of Materials Selection in Product Design*. Oxford: Butterworth–Heinemann.
Cooper, G. M. 2000. *The Cell: A Molecular Approach*. Washington, DC: ASM Press.
Courtney, T. H. 2000. *Mechanical Behavior of Materials*. Boston: McGraw–Hill.
Cullity, B. D. 1992. *Introduction to Magnetic Materials*. Reading, Mass.: Addison–Wesley.
Cullity, B. D., and S. R. Stock. 2001. *Elements of X-ray Diffraction*. Upper Saddle River, N.J.: Prentice Hall.
Grove, A. S. 1967. *Physics and Technology of Semiconductor Devices*. New York: John Wiley & Sons.
Halling, J. 1973. *Principles of Tribology*. New York: MacMillan.
Hoel, P. G. 1971. *Introduction to Mathematical Statistics*. New York: John Wiley & Sons.
Kittel, C. 2005. *Introduction to Solid State Physics*. Hoboken, N.J.: John Wiley & Sons.
Ohring, M. 2001. *Materials Science of Thin Films*. San Diego, Calif.: Academic Press.
Reed–Hill, R. E. 1973. *Physical Metallurgy Principles*. New York: van Nostrand.
Suresh, S. 1991. *Fatigue of Materials*. London: Cambridge University Press.

SPECIAL REFERENCE BOOKS

Amato, I. 1997. *Stuff, the Materials the World Is Made of*. New York: BasicBooks.
Dorr, R. F. 2002. *Boeing 747-400* (AirlinerTech Series, vol. 10). Specialty Press.
Glass, D. 2005. *Why Socks Disappear in the Wash*. New York: Barnes & Noble.
Haley, G., and R. Haley. 2000. *Haley's Cleaning Hints*. Hushion House Publishing.
Sass, S. L. 1998. *The Substance of Civilization*. New York: Arcade Publishing.

ARTICLES

Abramson, E. P. III. 1962. Dilute alloying effects on recrystallization in nickel as compared with other transition element solvents. *Transactions of the Metallurgical Society of AIME* 224:727.
Barsoum, M. W., E. N. Hoffman, R. D. Doherty, S. Gupta, and A. Zavaliangos. 2004. Driving force and mechanism for spontaneous metal whisker formation. *Physics Review Letters* 93:206104-1.
Frazer, B. W. 2000. The Gimli glider. *AOPA Pilot* July:74.
Fuchs, K. 1938. The conductivity of thin metallic films according to the electron theory of metals. *Proceedings of the Cambridge Philosophical Society* 34:100.
George, E. P., C. T. Liu, and D. P. Pope. 1992. Environmental embrittlement: The major cause of room-temperature brittleness in polycrystalline Ni_3Al. *Scripta Metallurgica* 27:365.
Hilliard, J. E. 1979. Phase diagrams and the engineer. *Journal of Educational Modules for Materials Science and Engineering* 1:173.
Johansson, C. H., and J. O. Linde. 1936. *Annals of Physics* 25:1
Lee, K. W., Y. H. Chen, Y. W. Chung, and L. M. Keer. 2004. Hardness, internal stress, and thermal stability of TiB_2/TiC multilayer coatings synthesized by magnetron sputtering with and without substrate rotation. *Surface and Coatings Technology* 177/178:591.
Li, D., Y. Chen, Y. W. Chung, and F. L. Freire, Jr. 2003. Metrology of 1–10 nm thick CN_x films: Thickness, density and surface roughness measurements. *Journal of Vacuum Science and Technology* A21:L19.
Linde, J. O. 1932. *Annals of Physics* 15:219.
Sondheimer, E. H. 1952. The mean free path of electrons in metals. *Advances in Physics* 1:1.

Index

A

Abrasive wear, 209
   coating hardness as figure of merit for, 261
Absolute temperature, 33, 44
Acceptor impurities, in semiconductors, 69
Activation barrier, 43
   and corrosion current, 198
Activation energy
   Arrhenius plot of, 57
   for creep, 103
   for diffusion along grain boundaries, 45
   and substrate-enzyme interactions, 113
   for vacancy formation, 43
Activation polarization, 198
Active sites, in polymer classification, 175
Adhesive wear, 209
Adiabatic demagnetization, 217
Aerodynamic drag, 90–91
Aerodynamics, specifications for Gossamer aircraft, 90–91
Air density, 90
Aircraft
   fatigue as mode of failure in, 110
   Gossamer aircraft specifications, 90–91
   strength-to-weight ratio, 89
   turbine blades in jet engines, 103
Airspeed, 90
AISI-SAE naming conventions, 135
Alkali halides, cohesive energies of, 7
Alkaline batteries, 204
Alkaline solutions, in corrosion, 189
Allotropes, 18
Alloys
   corrosion and oxidation, 187
   evaporation methods for thin films, 235–236
Alpha-helixes, in protein conformation, 138
Alternating copolymers, 177
Alternating current, converting to direct current, 209, 219
Alternators, 210, 227
Alumina
   as oxygen ionic conductor in hydrogen fuel cell, 193
   shock resistance relative to fused silica, 152
   thermal stress in, 151
Aluminum
   FCC structure, 16
   H/E ratio of, 262
Aluminum alloys, yield strength for, 95
Aluminum oxide
   insulation properties, 157
   as passivation layer, 199, 200
Amino acids, 137
Amino group, in amino acids, 137
Amonton's law, 257

Amorphous magnetic alloys, 222
Amorphous polymers, 177
Amorphous semiconductors, 42
Amorphous solids, 41–42
Amplified stress, near cracks/flaws, 108
Amps, 61
Annealing process, and defect removal, 105
Anodes
   and composition difference, 195
   in corrosion reactions, 187
   potential in stressed solids, 196
Antiferromagnetism, 218
Antiplane shear, 109
Antireflective coatings, 249, 250
Antisite defects, in ceramics, 147
Apparent saturation magnetization, 226
Applied field, 214
Applied stress, 103
Aramid, 178, 184
   chemical structure, 176
Archard's law, 260–261
Areal storage density
   50 million-fold increase in, 233
   increasing by reducing magnetic signal, 222
Argon
   ion bombardment energy, 242
   as sputter gas in titanium deposition, 240
Arrhenius plot, for processes with different activation energies, 57
Artificial knee joints, 1
Asperity contacts, in thin films, 260
Atomic bonding, 5–6
Atomic diffusion, 44–45
   along grain boundaries under stress, 103
   creep due to, 102
   and diffusion distance, 47–49
   due to step-function concentration profile, 47
   eliminating stored strain energy by, 105
   Fick's first law of diffusion, 45–46
   Fick's second law of diffusion, 46–47
Atomic plane stacking, 40
Atomic size, and mutual solubility, 35
Atomic weight, and corrosion rate, 197
Austenite, 132, 135
   in iron production, 130
Austenite/ferrite/cementite phase diagrams, 132
Automobile tires, copolymer composition, 177
Average grain size, 27
Aviation, 1
Avogadro's number, 190
Axial strain, 93
Axons, electrical signaling in, 82

Related Titles

The CRC Materials Science and Engineering Handbook, Third Edition
James F. Shackelford and William Alexander
ISBN: 0-8493-2699-6

Electronic, Magnetic and Optical Materials
Pradeep Fulay
ISBN: 0-8493-9564-X

Introduction to Engineering Materials, Second Edition
George Murray, Charles V. White, Mark Palmer, and Wolfgang Weise
ISBN: 1-57444-683-5

Introduction to the Principles of Materials Evaluation
David C. Jiles
ISBN: 0-8493-7392-1

Materials Processing Handbook
Joanna R. Groza, James F. Shackelford, and Enrique J. Lavernia
ISBN: 0-8493-3216-8

Materials Science for Engineers, Fifth Edition
J.C. Anderson, Keith D. Leaver, Rees D. Rawlings, and Patrick S. Leevers
ISBN: 0-7487-6365-1

Selection of Engineering Materials and Adhesives
Lawrence W. Fisher
ISBN: 0-8247-4047-5

STANDARD ATOMIC WEIGHTS (2001)

Z	Element	Symbol	Atomic Weight	Z	Element	Symbol	Atomic Weight
1	Hydrogen	H	1.00794(7)	58	Cerium	Ce	140.116(1)
2	Helium	He	4.002602(2)	59	Praseodymium	Pr	140.90765(2)
3	Lithium	Li	6.941(2)	60	Neodymium	Nd	144.24(3)
4	Beryllium	Be	9.012182(3)	61	Promethium	Pm	[144.9127]
5	Boron	B	10.811(7)	62	Samarium	Sm	150.36(3)
6	Carbon	C	12.0107(8)	63	Europium	Eu	151.964(1)
7	Nitrogen	N	14.0067(2)	64	Gadolinium	Gd	157.25(3)
8	Oxygen	O	15.9994(3)	65	Terbium	Tb	158.92534(2)
9	Fluorine	F	18.9984032(5)	66	Dysprosium	Dy	162.500(1)
10	Neon	Ne	20.1797(6)	67	Holmium	Ho	164.93032(2)
11	Sodium	Na	22.989770(2)	68	Erbium	Er	167.259(3)
12	Magnesium	Mg	24.3050(6)	69	Thulium	Tm	168.93421(2)
13	Aluminum	Al	26.981538(2)	70	Ytterbium	Yb	173.04(3)
14	Silicon	Si	28.0855(3)	71	Lutetium	Lu	174.967(1)
15	Phosphorus	P	30.973761(2)	72	Hafnium	Hf	178.49(2)
16	Sulfur	S	32.065(5)	73	Tantalum	Ta	180.9479(1)
17	Chlorine	Cl	35.453(2)	74	Tungsten	W	183.84(1)
18	Argon	Ar	39.948(1)	75	Rhenium	Re	186.207(1)
19	Potassium	K	39.0983(1)	76	Osmium	Os	190.23(3)
20	Calcium	Ca	40.078(4)	77	Iridium	Ir	192.217(3)
21	Scandium	Sc	44.955910(8)	78	Platinum	Pt	195.078(2)
22	Titanium	Ti	47.867(1)	79	Gold	Au	196.96655(2)
23	Vanadium	V	50.9415(1)	80	Mercury	Hg	200.59(2)
24	Chromium	Cr	51.9961(6)	81	Thallium	Tl	204.3833(2)
25	Manganese	Mn	54.938049(9)	82	Lead	Pb	207.2(1)
26	Iron	Fe	55.845(2)	83	Bismuth	Bi	208.98038(2)
27	Cobalt	Co	58.933200(9)	84	Polonium	Po	[208.9824]
28	Nickel	Ni	58.6934(2)	85	Astatine	At	[209.9871]
29	Copper	Cu	63.546(3)	86	Radon	Rn	[222.0176]
30	Zinc	Zn	65.409(4)	87	Francium	Fr	[223.0197]
31	Gallium	Ga	69.723(1)	88	Radium	Ra	[226.0254]
32	Germanium	Ge	72.64(1)	89	Actinium	Ac	[227.0277]
33	Arsenic	As	74.92160(2)	90	Thorium	Th	232.0381(1)
34	Selenium	Se	78.96(3)	91	Protactinium	Pa	231.03588(2)
35	Bromine	Br	79.904(1)	92	Uranium	U	238.02891(3)
36	Krypton	Kr	83.798(2)	93	Neptunium	Np	[237.0482]
37	Rubidium	Rb	85.4678(3)	94	Plutonium	Pu	[244.0642]
38	Strontium	Sr	87.62(1)	95	Americium	Am	[243.0614]
39	Yttrium	Y	88.90585(2)	96	Curium	Cm	[247.0704]
40	Zirconium	Zr	91.224(2)	97	Berkelium	Bk	[247.0703]
41	Niobium	Nb	92.90638(2)	98	Californium	Cf	[251.0796]
42	Molybdenum	Mo	95.94(2)	99	Einsteinium	Es	[252.0830]
43	Technetium	Tc	[97.9072]	100	Fermium	Fm	[257.0951]
44	Ruthenium	Ru	101.07(2)	101	Mendelevium	Md	[258.0984]
45	Rhodium	Rh	102.90550(2)	102	Nobelium	No	[259.1010]
46	Palladium	Pd	106.42(1)	103	Lawrencium	Lr	[262.1097]
47	Silver	Ag	107.8682(2)	104	Rutherfordium	Rf	[261.1088]
48	Cadmium	Cd	112.411(8)	105	Dubnium	Db	[262.1141]
49	Indium	In	114.818(3)	106	Seaborgium	Sg	[266.1219]
50	Tin	Sn	118.710(7)	107	Bohrium	Bh	[264.12]
51	Antimony	Sb	121.760(1)	108	Hassium	Hs	[277]
52	Tellurium	Te	127.60(3)	109	Meitnerium	Mt	[268.1388]
53	Iodine	I	126.90447(3)	110	Darmstadtium	Ds	[271]
54	Xenon	Xe	131.293(6)	111	Roentogenium	Rg	[272]
55	Cesium	Cs	132.90545(2)	112	Ununbium	Uub	[285]
56	Barium	Ba	137.327(7)	114	Ununquadium	Uuq	[289]
57	Lanthanum	La	138.9055(2)	116	Ununhexium	Uuh	[289]